PBY - Top Rated Western Adventures

Picked-By-You Guides®
Top Rated Outdoor Series

Top Rated™ Western Adventures

Guest Ranches, Pack Trips and Cattle Drives in North America

by Maurizio Valerio

PICKED-BY-YOU GUIDES®

Top Rated Outdoor Series

Copies of this book can be ordered from:
Picked-By-You
PO Box 718
Baker City, OR 97814
Phone: (800) 279-0479 • Fax: (541) 523-5028
www.topguides.com • e-mail: maurice@topguides.com

Artwork by Steamroller Studios, Cover Art by Fifth Street Design
Maps by Map Art, Cartesia Software
Cover vertical photo & thumbnail by Nine Quarter Circle Ranch,
Kim & Kelly Kelsey, Gallatin Gateway, Mt.
Printed in Korea

Publisher's Cataloging-in-Publication
(Provided by Quality Books, Inc.)

Valerio, Maurice.
Top rated Western adventures : guest ranches, pack trips and cattle drives in North America / by Maurizio Valerio. -- 1st ed.
p. cm. -- (Top rated outdoor series)
Includes indexes.
Preassigned LCCN: 98-67997
ISBN: 1-889807-08-7

1. Dude ranches--West (U.S.)--Directories. 2. Cattle drives--West (U.S.)--Directories. 3. Trail riding--West (U.S.)--Directories. I. Title.

GV198.96.W47V35 1999 796.56'0258
QBI98-1369

Please note: Picked-By-You LLC and the author have used their best efforts in preparing this book, but they do not intend to portray this book as an all-encompassing authority on the outdoors. This book was developed as a consumer guide for the outdoor enthusiast. All outdoor professionals and businesses featured in this book were selected by their past clients, according to the rules explained in the book. Picked-By-You LLC and the author make no representation with regard to the past and/or future performance of the outdoor professionals and businesses listed. Picked-By-You LLC and the author make no representation or warranties with respect to the accuracy or completeness of the contents of this book, and specifically disclaim any implied warranties, merchantability, or fitness for any particular purpose and shall in no event be liable for any injury, loss of profit or any other commercial damage, including but not limited to special, incidental, consequential, or other damages.

To Allison, Marco and Nini

About the Author

Maurizio (Maurice) Valerio received a Doctoral degree Summa cum Laude in Natural Science, majoring in Animal Behaviour, from the University of Parma (Italy) in 1981, and a Master of Arts degree in Zoology from the University of California, Berkeley in 1984.

He is a rancher, a writer and a devoted outdoorsman who decided to live with the wild animals that he cherishes so much in the Wallowa Mountains of Northeast Oregon. He has traveled extensively in the Old and New World, for more than 25 years. He is dedicated to preserving everyone's individual right of a respectful, knowledgeable and diversified use of our Outdoor Resources.

Table of Contents

Acknowledgments

It is customary in this section to give credit to those who have contributed to the realization of the end product. The Picked-By-You Guides® started three years ago as a little personal crusade and has evolved into a totally challenging, stimulating and rewarding full time commitment.

My deep thanks must go first to all the Captains, Ranchers, Guides, Lodges and Outfitters who decided to trust our honesty and integrity. They have taken a leap of faith in sharing their lists of clients with us and for this we are truly honored and thankful.

They have constantly encouraged our idea. Captains have taught us the difference between skinny fishing and skinny dipping, while River Guides have patiently help us to identify rafters , purlins , catarafts and J-rig rafts. They were also ready to give us a badly needed push forward every time this very time-consuming idea came to a stall. We have come to know many of them through pleasant phone chats, e-mails, faxes and letters. They now sound like old friends on the phone and we are certain we all share a deep respect for the mountains, the deserts and the waters of this great country of ours.

The Picked-By-You Team (both in the office and abroad), with months of hard work, skills, ingenuity, good sense of humor and pride, have then transformed a simple good idea into something a bit more tangible and much more exciting. They all have put their hearts in the concept and their hands and feet in the dirt. Some with a full-time schedule, some with a part-time collaboration, all of them bring their unique and invaluable style and contribution.

My true thanks to Brent Beck, Lindsay Benson, Bob Erdmann, Robert Evans, Cheryl Fisher, Brian Florence, Sally Georgeson, Grace Martin, Kevin McNamara, Jerry Meek, Allison C. Mickens, Tom Novak, Shelby Sherrod, Dyanne Van Swoll, Giuseppe Verdi and Mr. Peet's Coffee and Tea.

Last, but not least, my sincere, profound, and loving gratitude to my wife Allison. Her patient support, her understanding, her help and her skills have been the fuel which started and stoked this fire. Her laughter has been the wind to fan it.

To you Allie, with a toast to the next project…just kidding!

Maurizio (Maurice) Valerio

Preface

The value of information depends on its usefulness. Simply put, whatever allows you to make informed choices will be to your advantage. To that end, Picked-By-You Guides® aims to take the guesswork out of selecting services for outdoor activities. Did you get what you paid for? From Picked-By-You Guides®' point of view, the most reliable indicator is customer satisfaction.

The information in this book is as reliable as those who chose to participate. In the process of selecting the top professionals, Picked-By-You Guides® contacted all licensed guides, outfitters and businesses which provide services for outdoor activities. They sought to include everyone but not all who were contacted agreed to participate according to the rules. Thus, the omission of a guide, outfitter or service does not automatically mean they didn't qualify based on customer dissatisfaction.

The market abounds with guidebooks by 'experts' who rate a wide range of services based on their personal preferences. The value of the Picked-By-You concept is that businesses earn a place in these books only when they receive favorable ratings from a majority of clients. If ninety percent of the customers agree that their purchase of services met or exceeded their expectations, then it's realistic to assume that you will also be satisfied when you purchase services from the outdoor professionals and businesses included in this book.

It's a fact of life; not everyone is satisfied all of the time or by the same thing. Individual experiences are highly subjective and are quite often based on expectations. One person's favorable response to a situation might provoke the opposite reaction in another. A novice might be open to any experience without any preconceived notions while a veteran will be disappointed when anything less than great expectations aren't met.

If you select any of the businesses in this book, chances are excellent that you will know what you are buying. A diversity of clients endorsed them because they believed the services they received met or exceeded their expectations. Picked-By-You Guides® regards that information more valuable than a single observer or expert's point of view.

The intent behind Picked-By-You Guides® is to protect the consumer from being misled or deceived. It is obvious that these clients were given accurate information which resulted in a positive experience and a top rating.

The number of questionnaire responses which included detailed and sometimes lengthy comments impressed upon us the degree to which people value their experiences. Many regard them as "once-in-a-lifetime" and "priceless," and they heaped generous praise on those whose services made it possible.

Picked-By-You Guides® has quantified the value of customer satisfaction and created a greater awareness of top-rated outdoor professionals. It remains up to you to choose and be the judge of your own experience. With the help of this book, you will have the advantage of being better informed when making that pick.

Robert Evans, *information specialist*

The Picked-By-You Guides® Idea

Mission Statement

The intent of this publication is to provide the outdoor enthusiast and his/her family with an objective and easy-to-read reference source that would list only those businesses and outdoor professionals who have **agreed to be rated** and have been overwhelmingly endorsed by their past clients.

There are many great outdoor professionals (Guides, Captains, Ranches, Lodges, Outfitters) who deserve full recognition for putting their experience, knowledge, long hours, and big heart, into this difficult job. With this book we want to reward those deserving professionals while providing an invaluable tool to the general public.

Picked-By-You Guides® are the only consumer guides to outdoor activities.

In this respect it would be useful to share the philosophy of our Company succinctly illustrated by our Mission Statement:

> "To encourage and promote the highest professional and ethical standards among those individuals, Companies, Groups or Organizations who provide services to the Outdoor Community.

To communicate and share the findings and values of our research and surveys to the public and other key groups.

To preserve everyone's individual right of a respectful, knowledgeable and diversified use of our Outdoor Resources".

Our business niche is well defined and our job is simply to listen carefully.

THEY 'the experts' Vs. WE 'the People'

Picked-By-You books were researched and compiled by **asking people such as yourself**, who rafted, fished, hunted or rode a horse on a pack trip with a particular outdoor professional or business, to rate their services, knowledge, skills and performance.

Only the ones who received A- to A+ scores from their clients are found listed in these pages.

The market is flooded with various publications written by 'experts' claiming to be the ultimate source of information for your vacation. We read books with titles such as " The Greatest River Guides", "The Complete Guide to the Greatest Fishing Lodges" etc.

We do not claim to be experts in any given field, but we rather pass to history as good....listeners. In the preparation of the Questionnaires we listened first to the outdoor professionals' point of view and then to the comments and opinions of thousands of outdoor enthusiasts. We then organized the findings of our research and surveys in this and other publications of this series.

Thus we will not attempt to tell how to fish, how to paddle or what to bring on your trip. We are leaving this to the outdoor professionals featured in this book, for they have proven to be outstanding in providing much valuable information before, during and after your trip.

True [paid] advertising: an oxymoron

Chili with beans is considered a redundant statement for the overwhelming majority of cooks but it is an insulting oxymoron for any native Texan.

In the same way while 'true paid advertising' is a correct statement for some, it is a clear contradiction in terms for us and certainly many of you. A classic oxymoron.

This is why we do not accept commissions, donations, invitations, or, as many publishers cleverly express it, "...extra fees to help defray the cost of publication". Many articles are written every month in numerous specialized magazines in which the authors tour the country from lodge to lodge and camp to camp sponsored, invited, or otherwise compensated in many different shapes or forms.

It is indeed a form of direct advertising and, although this type of writing usually conveys a good amount of general information, in most cases it lacks the impartiality so valuable when it comes time to make the final selection for your vacation or outdoor adventure.

Without belittling the invaluable job of the professional writers and their integrity, we decided to approach the task of **researching information and sharing it with the public** with a different angle and from an opposite direction.

Money? .. No thanks!

We are firmly **committed to preserve the impartiality** and the novelty of the Picked-By-You idea.

For this reason we want to reassure the reader that the outdoor professionals and businesses featured in this book have not paid (nor will they pay), any remuneration to Picked-by-You Guides® or the author in the form of money, invitations or any other considerations.

They have earned a valued page in this book solely as the result of ***their hard work and dedication to their clients.***

"A spot in this book cannot be purchased: it must be earned"

Size of a business in not a function of its performance

Since the embryonic stage of the Picked-By-You idea, during the compilation of the first Picked-By-You book, we faced a puzzling dilemma.

Should we establish a minimum number of clients under which a business or outdoor professional will not be allowed to participate in our evaluating process?

This would be a 'safe' decision when it comes the time to elaborate the responses of the questionnaires. But we quickly learned that many outdoor professionals limit, by choice, the total number of clients and, by philosophy of life, contain and control the size of their business. They do not want to grow too big and sacrifice the personal touches or the freshness of their services. In their words "we don't want to take the chance to get burned out by people." They do not consider their activity just a job, but rather a way of living.

"WHY, NO MAM, WE NEVER HAVE HAD ANY OF THOSE SASQUATCH SIGHTINGS IN THESE PARTS."

But if this approach greatly limits the number of clients accepted every year we must say that these outdoor professionals are the ones who often receive outstanding ratings and truly touching comments from their past clients.

Some businesses have provided us with a list of clients of 40,000, some with 25 . In this book **you will find both the large and the small**.

From a statistical point, it is obvious that a fly fishing guide who submitted a list of 32 clients, by virtue of the sample size of the individuals surveyed, will implicitly have a lower level of accuracy if compared to a business for which we surveyed 300 guests. (Please refer to the Rating and Data

Elaboration Sections for details on how we established the rules for qualification and thus operated our selection).

We do not believe that the size of business is a function of its good performance and we feel strongly that those dedicated professionals who choose to remain small deserve an equal chance to be included.

We tip our hats

. We want to recognize all the Guides, Captains, Ranches, Lodges and Outfitters who have participated in our endeavor, whether they qualified or not. The fact alone that they accepted to be rated by their past clients is a clear indication of how much they care, and how willing they are to make changes.

We also want to credit all those outdoor enthusiasts who have taken the time to complete the questionnaires and share their memories and impressions with us and thus with you. Some of the comments sent to us were hilarious, some were truly touching.

We were immensely pleased by the reaction of the outdoor community at large. The idea of "Picked-by-You Guides®" was supported from the beginning by serious professionals and outdoor enthusiasts alike. We listened to their suggestions, their comments, their criticisms and we are now happy to share this information with you.

Questionnaires

"Our books will be only as good as the questions we ask."

We posted this phrase in the office as a reminder of the importance of the 'tool' of this trade. The questions.

Specific Questionnaires were tailored to each one of the different activities surveyed for this series of books. While a few of the general questions remained the same throughout, many were specific to particular activities. The final objective of the questionnaire was to probe the many different facets of that diversified field known as the outdoors.

The first important factor we had to consider in the preparation of the Questionnaires was the total number of questions to be asked. Research shows an *inversely proportionate relation* between the total number of questions and the percentage of the response: the higher the number of questions, the lower the level of response. Thus we had to balance an acceptable return rate with a meaningful significance. We settled for a compromise and we decided to keep 20 as the maximum number.

The first and the final versions of the Questionnaires on which we based our surveys turned out to be very different. We asked all the businesses and outdoor professionals we contacted for suggestions and criticisms. They helped us a great deal: we weighed their different points of view and we incorporated all their suggestions into the final versions.

We initially considered using a phone survey, but we quickly agreed with the businesses and outdoor professional that we all are already bothered by too many solicitation calls when we are trying to have a quiet dinner at home. We do not want you to add Picked-By-You to the list of companies that you do not want to talk to, nor we want you to add our 800 number to your caller ID black list.

In using the mail we knew that we were going to have a slightly lower percentage of questionnaires answered, but this method is, in our opinion, a more respectful one.

We also encouraged the public to participate in the designing of the questionnaire by posting on our Web at www.topguides.com the opportunity to submit a question and ...."Win a book". Many sent their suggestions and , if they were chosen to be used in one of our questionnaires, they were given the book of their choice.

Please send us your question and/or your suggestions for our future surveys at:

PICKED-BY-YOU Guides®, P.O. Box 718, Baker City, OR 97814

Rating (there is more than one way to skin the cat)

We considered many different ways to score the questionnaires, keeping in mind at all times our task:

translate an opinion into a numerical value

Some of the approaches considered were simple *averages* [arithmetical means], others were sophisticated statistical tests. In the end we opted for simplicity, sacrificing to the God of statistical significance. WARNING: if $p \leq 0.001$ has any meaning in your life stop reading right here: you will be disappointed with the rest.

For the rest of us, we also made extensive use in our computation of the *median*, a statistic of location, which divides the frequency distribution of a set of data into two halves. A quick example, with our imaginary Happy Goose Outfitter, will illustrate how in many instances the *median* value, being the center observation, helps describing the distribution, which is the truly weak point of the *average*:

Average salary at Happy Goose Outfitters $ 21,571

Median salary at Happy Goose Outfitters $ 11,000

5,000	10,000	10,000	11,000	12,000	15,000	98,000
Wrangler	Guide	Guide	Senior Guide	Asst.Cook	Cook	Boss

Do not ask the boss : "What's the average salary?"

These are the values assigned to **Questions 1-15**:

5.00 points	OUTSTANDING
4.75 points	EXCELLENT
4.25 points	GOOD
3.50 points	ACCEPTABLE
3.00 points	POOR
0.00 points	UNACCEPTABLE

Question 16, relating to the weather conditions, was treated as bonus points to be added to the final score.

Good=0 Fair=1 Poor=2

The intention here was to reward the outdoor professional who had to work in adverse weather conditions.

Questions 17 - 18 = 5 points

Questions 19 - 20 = 10 points

The individual scores of each Questionnaire were expressed as a percentage to avoid the total score from being drastically affected by one question left unanswered or marked "not applicable." All the scores received for each individual outdoor professional and business were thus added and computed.

The 90 points were considered our cutoff point. Note how the outfitters must receive a combination of Excellent with only a few Good marks (or better) in order to qualify.

Only the Outfitters, Captains, Lodges, Guides who received an A- to A+ score did qualify and they are featured in this book.

We also decided not to report in the book pages the final scores with which the businesses and the outdoor professionals ultimately qualified. In a way we thought that this could be distractive.

In the end, we must admit, it was tough to leave out some outfitters who scored very close to the cutoff mark.

It would be presumptuous to think that our scoring system will please everybody, but we want to assure the reader that we tested different computations of the data. We feel the system that we have chosen respects the overall opinion of the guest/client and maintains a more than acceptable level of accuracy.

We know that .... "You can change without improving, but you cannot improve without changing."

The Power of Graphs (how to lie by telling the scientific truth)

The following examples illustrate the sensational (and unethical) way with which the 'scientific' computation of data can be distorted to suit one's needs or goals.

The *Herald* presents a feature article on the drastic increase of total tonnage of honey stolen by bears (mostly Poohs) in a given area during 1997.

Total tonnage of honey stolen by bears (Poohs)

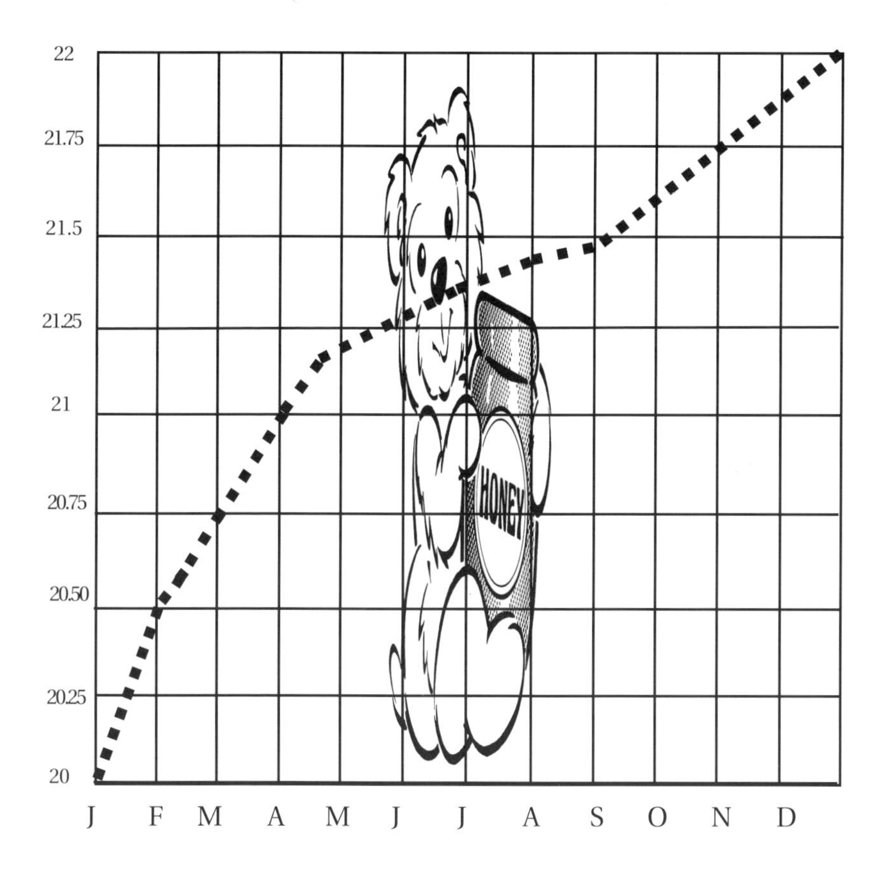

Total tonnage of honey stolen by bears (Poohs)

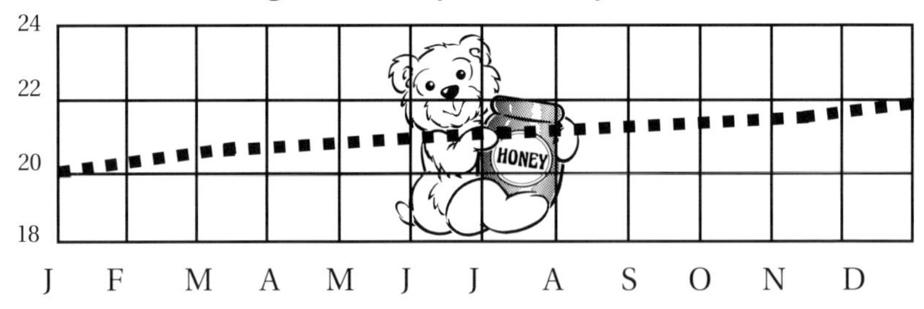

It is clear how a journalist, researcher or author must ultimately choose one type of graph. But the question here is whether or not he/she is trying to make "his/her point" by choosing one type versus the other, rather than simply communicate some findings.

Please note that the bears, in our example, are shameless, and remain such in both instances, for they truly love honey!

Graphs were not used in this book. We were just too worried we wouldn't use the correct ones.

The Book Making Process

Research

We **researched** the name and address of every business and outdoor professional **in the United States and** in all the **provinces of Canada** (see list in the Appendix). Some states do not require guides and outfitters or businesses providing outdoor services to be registered, and in these instances the information must be obtained from many different sources [Outfitter's Associations, Marine Fisheries, Dept. of Tourism, Dept. Environmental Conservation, Dept. of Natural Resources, Dept. of Fish and Game, US Coast Guard, Chamber of Commerce, etc.].

In the end the database on which we based this series of Picked-By-You Guides® amounted to more than 23,000 names of Outfitters, Guides, Ranches, Captains etc. Our research continues and this number is increasing every day. The Appendix in the back of this book is only a partial list and refers specifically to Top Rated Western Adventures.

Participation

We **invited** businesses and outdoor professionals, with a letter and a brochure explaining the Picked-By-You concept, to join our endeavor by simply sending us a **complete list of their clients** of the past two years. With the "Confidentiality Statement" we reassured them that the list was going to be kept **absolutely confidential** and to be *used one time only* for the specific purpose of evaluating their operation. Then it would be destroyed.

We truly oppose this "black market" of names so abused by the mail marketing business. If you are ever contacted by Picked-By-You you may rest assured that your name, referred to us by your outdoor professional, will never be sold, traded or otherwise used a second time by us for marketing purposes.

Questionnaires

We then **sent a questionnaire** to **every single client on each list** (to a maximum of 300 randomly picked for those who submitted large lists with priority given to overnight or multiple day trips), asking them to rate the

services, the **knowledge** and **performance** of the business or outdoor professional by completing our comprehensive questionnaire (see pages 146-147). The businesses and outdoor professionals found in these pages may or may not be the ones who invest large sums of money to advertise in magazines, or to participate at the annual conventions of different clubs and foundations. However, they are clearly the ones, according to our survey, that put customer satisfaction and true dedication to their clients first and foremost.

Data Elaboration

A **numerical value was assigned to each question**. All the **scores were computed**. Both the **average** and the **median** were calculated and considered for eligibility. Please note that the total score was computed as a percentile value.

This allows some flexibility where one question was left unanswered or was answered with a N/A. Furthermore, we decided not to consider the high and the low score to ensure a more evenly distributed representation and to reduce the influence which an extreme judgement could have either way (especially with the small sample sizes).

We also set a **minimum number of questionnaires** which needed to be answered to allow a business or an outdoor professional to qualify. Such number was set as a function of the total number of clients in the list: the smaller the list of clients, the higher was the percentage of responses needed for qualification.

In some cases the outfitter's average score came within 1 points of the A-cutoff mark. In these instances, we considered both the median and the average were considered as well as the guests' comments and the total number of times that this particular business was recommended by the clients by answering with a 'yes' question 19 and 20.

Sharing the results

Picked-By-You will share the results of this survey with the businesses and the outdoor professionals. This will be done at no cost to them whether or not they qualified for publication. All questionnaires received will, in fact, be returned along with a summary result to the business, keeping the confidentiality of the client's name when this was requested. This will prove an invaluable tool to help improving those areas that have received some criticisms.

The intention of this series of books is to research the opinions and the comments of outdoor enthusiasts, and to share the results of our research with the public and other key groups.

One outfitter wrote us about our Picked-by-You Guides® series, "I feel your idea is an exciting and unique concept. Hopefully our past clientele will rate us with enough points to 'earn' a spot in your publication. If not, could we please get a copy of our points/questionnaires to see where we need to improve. Sincerely..."

This outfitter failed to qualify by just a few points, but such willingness to improve leaves us no doubt that his/her name will be one of those featured in our second edition. In the end it was not easy to exclude some of them from publication, but we are certain that, with the feedback provided by this survey, they will be able to improve those areas that need extra attention.

We made a real effort to keep a position of absolute impartiality in this process and, in this respect, we would like to repeat that the outfitters have not paid, nor they will pay, one single penny to Picked-By-You Guides® or the Author to be included in this book.

The research continues.

Icon Legend

General Services and Accommodations

Infant

Toddler

Kids

Babysitter

Family

Full Board

Senior Citizen

Handicap

Natural /Gourmet Meals

Bed and Breakfast

Women Only Camps/Dates

Drop Camp

Icon Legend

General Services and Accommodations

Lodge

Dome / Spike Tent

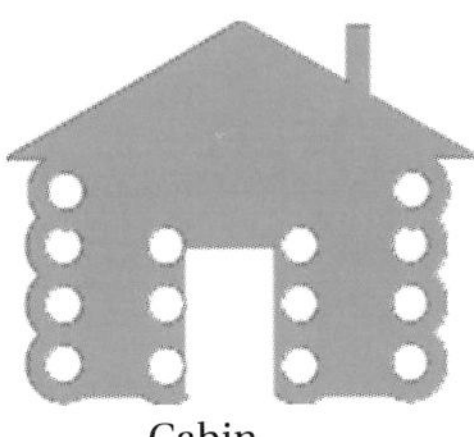

Cabin

Wall Tent

Hotel/Motel

Photography

Boots Provided

Supervised Sports

Trailer

Hot Springs/Spas

Archeological Sites

Swimming Pool

Dog Kennels

Racquetball

Tennis

Icon Legend

General Services

Unguided Activities

Guided Activities

Overnight Trips

Day Trips

Season(s) of Operation

Fall

Summer

Year-round

Winter

Spring

Icon Legend

Activities

Horseback Riding

Llama Pack Trips

Pack Trips

Barrel Racing

Roping

Team Roping

Team Penning

Steer Wrestling

Wagon Rides

Ghost Town Tours

Branding

Cutting

Cattle / Horse Drives

Icon Legend

Activities

Catch and Release

Fly Fishing

Fishing

Ice Fishing

Bow Fishing

Float Fishing

ATV Riding

Float Trips

Swimming

Snowmobiling

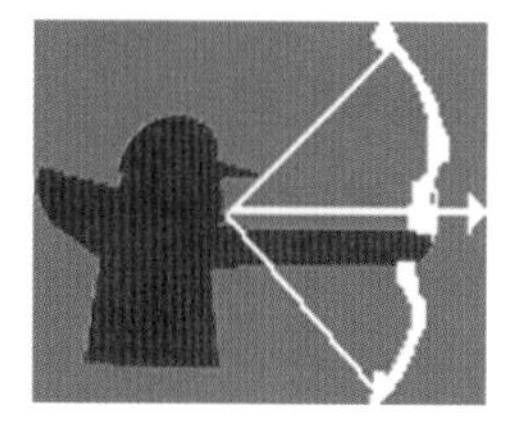

Bow Hunting/Archery

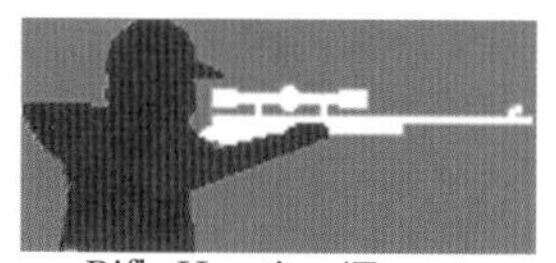

Rifle Hunting/Target Shooting

Bird Hunting

Big Game Hunting

Clay Shooting

Icon Legend

Activities

Hiking

Wildlife Observation

Backpack / Trekking Excursions

Dancing

Cowboy and Live Entertainment

Golfing

Crosscountry Skiing

Mountain Biking

Snowshoeing

Horse Shoeing School

Fly Fishing School

Horse Packing School

Horse Riding School

Western Adventures

Ranches, Outfitters and Pack Stations

Picked-By-You Professionals in

Alaska

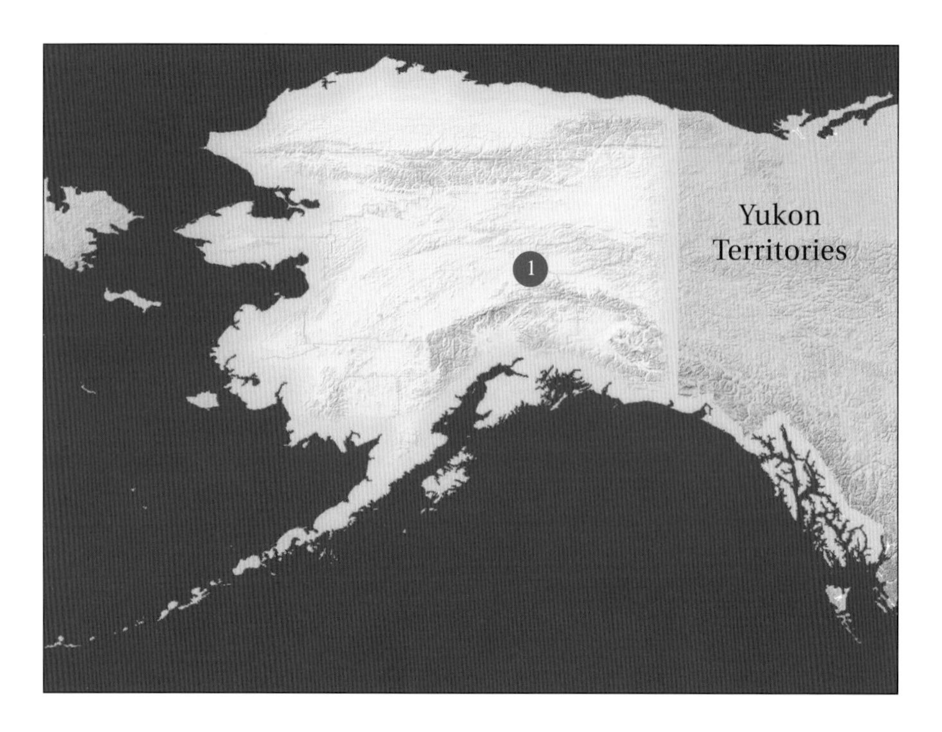

Outdoor Professionals

1 Lost Creek Ranch

Useful information for the state of

Alaska

State and Federal Agencies

Alaska Dept. of Fish & Game
PO Box 25556
Juneau, AK 99802-5526
phone: (907) 465-4100

Alaska Region Forest Service
709 West 9th Street
Box 21628
Juneau, AK 99802-1628
phone: (907) 586-8863
TTY: (907) 586-7816

Chugach National Forest
3301 C Street, Ste. 300
Anchorage, AK 99503-3998
phone: (907) 271-2500
TTY: (907) 271-2332

Tongass National Forest:
Chatham Area
204 Siginaka Way
Sitka, AK 99835
phone: (907) 747-6671
TTY: (907) 747-8840

Bureau of Land Management
Alaska State Office
222 W. 7th Avenue, #13
Anchorage, AK 99513-7599
phone: (907) 271-5960
or 907-271- Plus Extension
fax: (907) 271-4596

Office Hours: 7:30 a.m. - 4:15 p.m.

National Parks

Denali National Park
phone: (907) 683-2294

Gates of the Arctic National Park
phone: (907) 456-0281

Glacier Bay National Park
phone: (907) 697-2230

Katmai National Park
phone: (907) 246-3305

Kenai Fjords National Park
phone: (907) 224-3175

Kobuk Valley National Park
phone: (907) 442-3890

Lake Clark National Park
phone: (907) 271-3751

Wrangell-St. Elias National Park
phone: (909) 822-5235

Associations, Publications, etc.

DudeRanches.com
http://www.duderanches.com

License and Report Requirements

- State requires licensing of Outdoor Professionals.
- State requires a "Hunt Record" for big game.
- State to implement a "logbook" program for charter vessel/guided catches of King Salmon in Southeast Alaska by the 1998 season.

Lost Creek Ranch

Les and Norma Cobb
P.O. Box 84334 • Fairbanks, AK 99708
phone: (907) 672-3999

Lost Creek Ranch is located 150 miles northwest of Fairbanks, Alaska. Les Cobb has lived and hunted the area for 25 years.

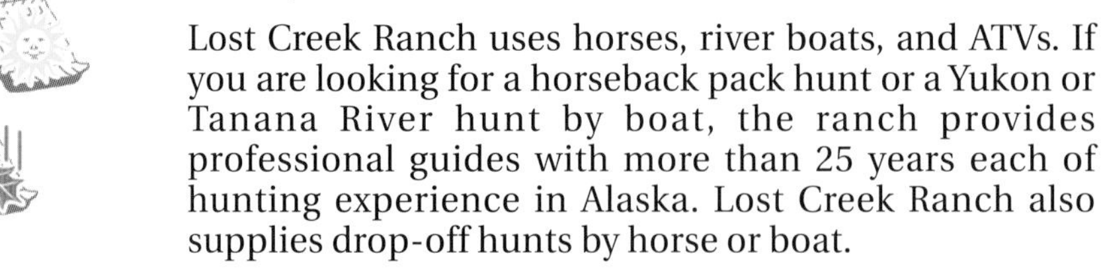

Lost Creek Ranch uses horses, river boats, and ATVs. If you are looking for a horseback pack hunt or a Yukon or Tanana River hunt by boat, the ranch provides professional guides with more than 25 years each of hunting experience in Alaska. Lost Creek Ranch also supplies drop-off hunts by horse or boat.

Since it is a guest ranch, there is always something for the whole family to enjoy together, such as trail riding, fishing, nature hikes, or the pioneer flavor of life with peace and quiet all around.

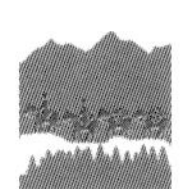

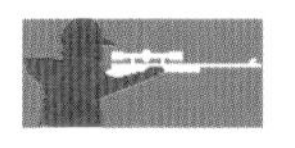

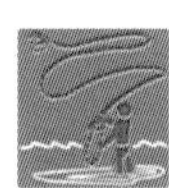

"Les & Norma went out of their way with us and other guests to make our Alaskan trip an Alaskan adventure!" Peter Burokas

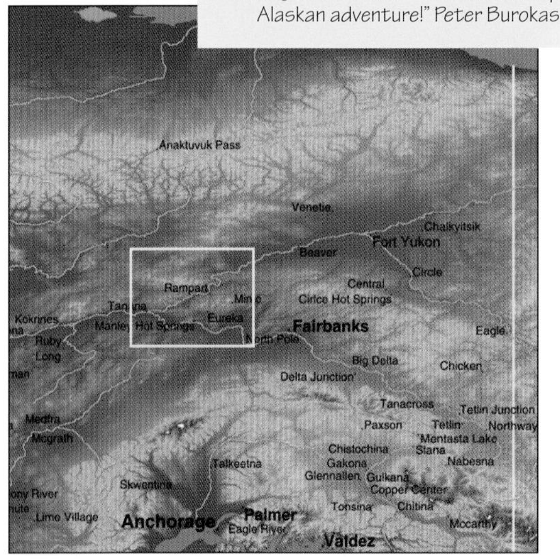

Lost Creek Ranch

Picked-By-You Professionals in

California

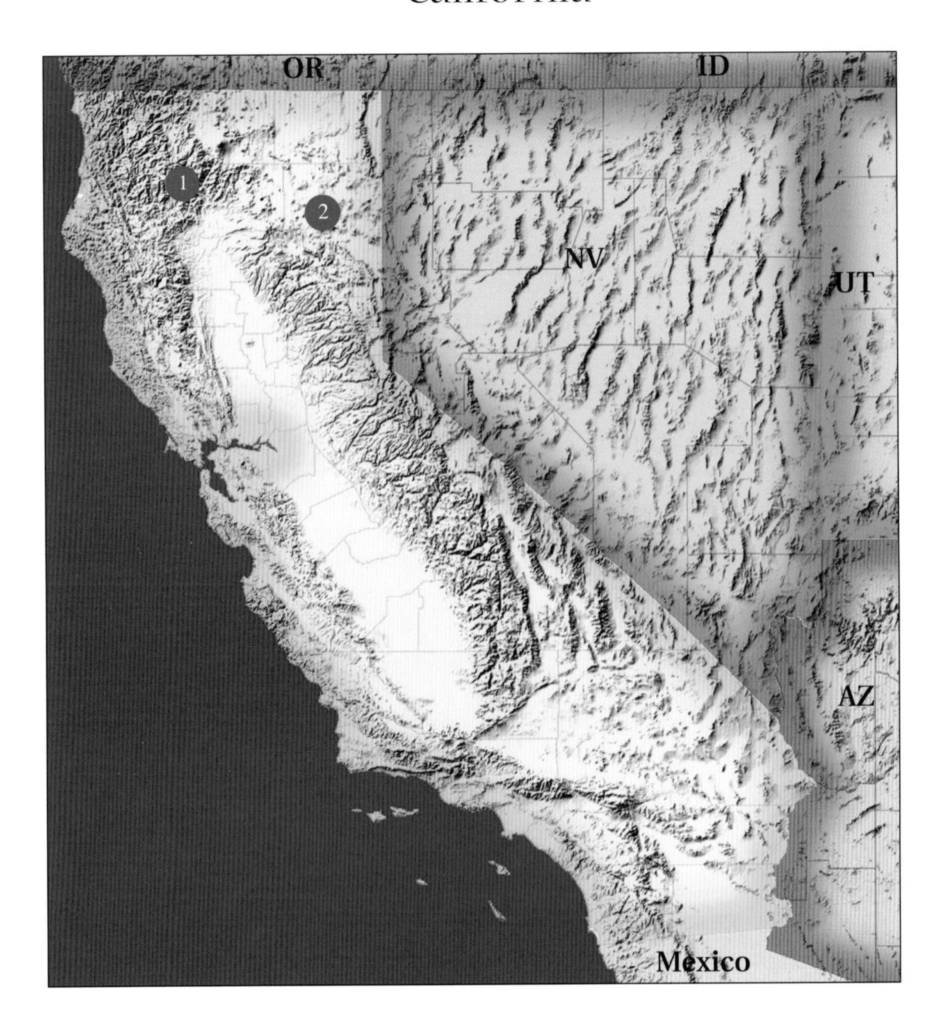

Outdoor Professionals

1. Coffee Creek Ranch
2. Spanish Springs Ranch

Useful information for the state of

California

State and Federal Agencies

California Fish & Game Commission
License & Revenue Branch
1416 9th Street
Sacramento, CA 95814
phone: (916) 227-2244

Pacific Southwest
Forest Service Region
630 Sansome St.
San Francisco, CA 94111
phone: (415) 705-2874
TTY: (415) 705-1098

Inyo National Forest
phone: (619) 373-2400

Klamath National Forest
phone: (916) 842-6131

Lake Tahoe Basin
phone: (916) 573-2600

Lassen National Forest
phone: (916) 257-2151

Modoc National Forest
phone: (916) 233-5811

Sequoia National Forest
phone: (209) 784-1500

Shasta Trinity National Forest
phone: (916) 246-5222

Sierra National Forest
phone: (209) 297-0706

Stanislaus National Forest
phone: (209) 532-3671

Tahoe National Forest
phone: (916) 265-4531

Bureau of Land Management
California State Office
2135 Butano Drive
Sacramento, CA 95825
phone: (916) 978-4400
or (916) 978-Plus Extension
fax: (916) 978-4620
Office Hours: 7:30 - 4:00 p.m. (PST)

National Parks

Lassen Volcanic National Park
phone: (916) 595-4444

Redwood NationalPark
phone: (707) 464-6101

Sequoia & Kings Canyon National Parks
phone: (209) 565-3341

Yosemite National Park
phone: (209) 372-0200

Channel Islands National Park
phone: (805) 658-5700

Associations, Publications, etc.

California Outdoors
PO Box 401
Coloma, CA 95613
phone: (800) 552-3625

DudeRanches.com
http://www.duderanches.com

License and Report Requirements

- State requires licensing of Outdoor Professionals.
- State requires the filing of a "Monthly Guide Log" for fishing and hunting.
- River Outfitters need a "Use Permit", required for BLM, National Forest, Indian reservations, and National Parks

Coffee Creek

Ruth Hartman
HC 2, Box 4940 – Dept AW • Trinity Center, CA 96091
phone: (800) 624-4480 • fax: (916) 266-3597

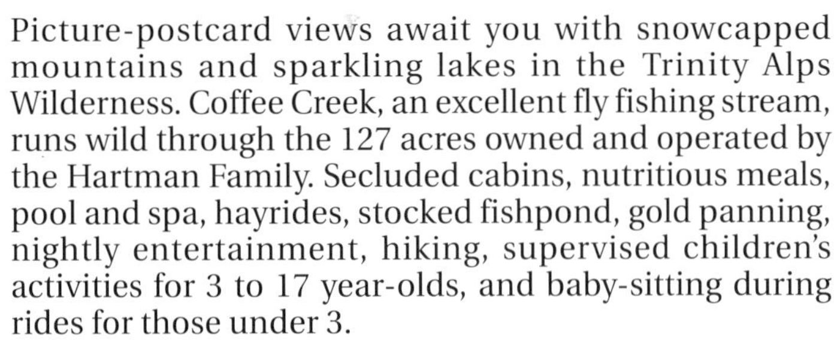

Picture-postcard views await you with snowcapped mountains and sparkling lakes in the Trinity Alps Wilderness. Coffee Creek, an excellent fly fishing stream, runs wild through the 127 acres owned and operated by the Hartman Family. Secluded cabins, nutritious meals, pool and spa, hayrides, stocked fishpond, gold panning, nightly entertainment, hiking, supervised children's activities for 3 to 17 year-olds, and baby-sitting during rides for those under 3.

Scenic mountain trails, some roping and horsemanship lessons, and gymkhana. Spring, fall and senior discounts. Romantic weekends. Fall foliage.

Hunting for deer and bear and pack trips. Adult-only weeks. Wedding and reception packages, maximum 250 persons; small meetings-conferences, maximum 50. Winter cross country skiing, horse-drawn sleigh rides. Pickup service available from Redding, free for Trinity Center airport (private planes).

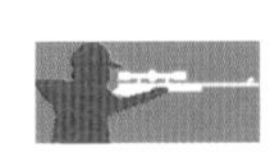

"The cabin accommodations are set next to the creek, the sound was so relaxing. The people have a yesteryear's friendly spirit and 'aim to please' attitude. I recommend Coffee Creek to anyone searching for a vacation with western flavor" Marilyn McIntosh

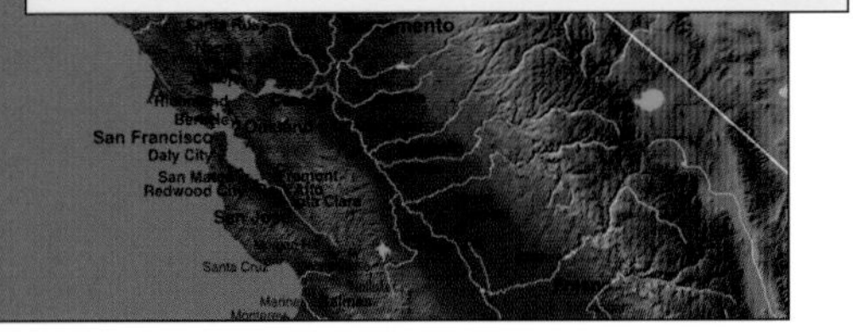

NORTHERN CALIFORNIA'S
FINEST
GUEST RANCH
NH
coffee
creek

Spanish Springs Ranch

Jim Vondracek & Sharon Roberts
P.O. Box 70 • Ravendale, CA 96123
phone: (800) 560-1900 • ranch (916) 234-2050 • fax: (916) 234-2041
email: spanishsprings@mcwaters.net • http://www.spanish-springs.com

Spanish Springs Ranch provides guests with authentic Western vacations. This working ranch of 70,000 pristine acres make for some of the most breathtaking horseback riding in the world. The ranch also features 4,000 cattle and more than 100 horses.

Activities include horseback riding, swimming, trap shooting, roping and much more.

The ranch is open from March through December. We offer comfortable facilities with all the amenities, and three hearty meals a day. Our goal is to provide each guest with the most enjoyable vacation possible.

There are always many great events happening at Spanish Springs Ranch. As you know, any time is a great time to come up and enjoy the Ranch.

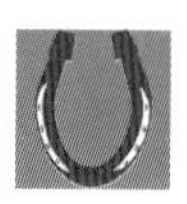

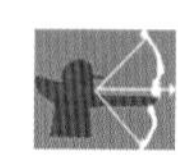

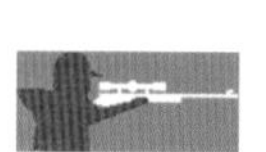

"His (Jim Vondracek) thorough knowledge of the region and attention to safety offer the right mix of excitement and relaxation. At the end of each day the chuck wagon is miraculously transformed into one of the finest open air cafes west of the Rockies!" Alex Estreicher

SPANISH SPRINGS
R·A·N·C·H
You Have Arrived
SPANISH SPRINGS

Picked-By-You Professionals in

Colorado

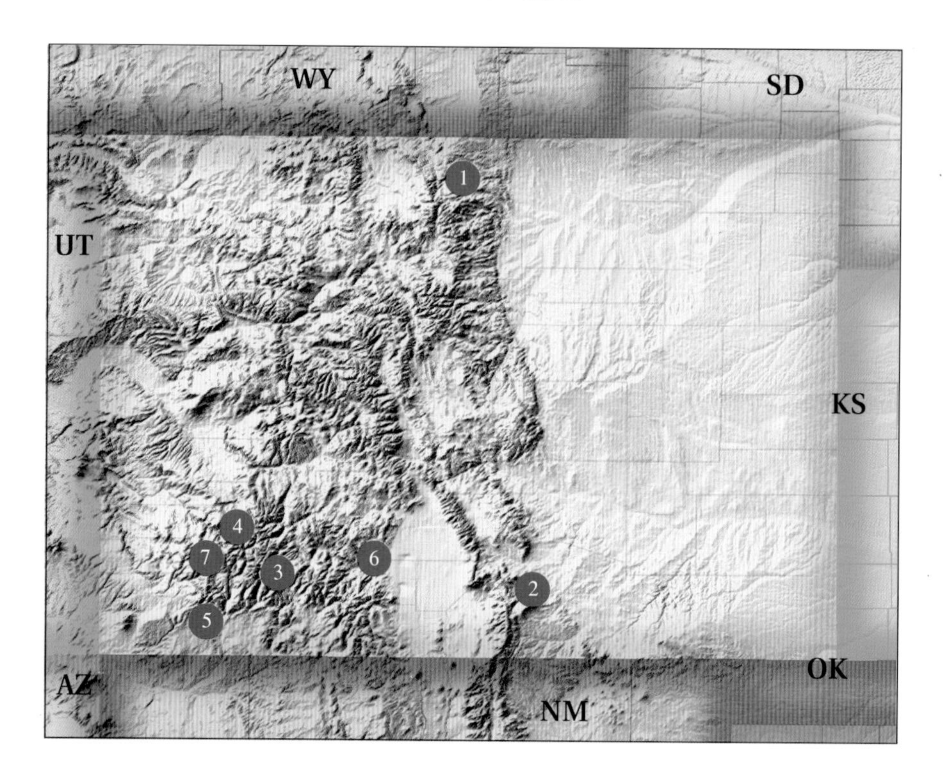

Outdoor Professionals

1. Beaver Meadows Resort Ranch
2. Echo Canyon Guest Ranch
3. Frazier Outfitting
4. Lakeview Resort & Outfitters
5. San Juan Outfitting
6. Schmittel Packing & Outfitting
7. Skyline Guest Ranch

Useful information for the state of

Colorado

State and Federal Agencies

Colorado Agencies of Outfitters Registry
1560 Broadway, Suite 1340
Denver, CO 80202
phone: (303) 894-7778

Colorado Dept. of Natural Resources
1313 Sherman, Room 718
Denver, CO 80203
phone: (303) 866-3311

Forest Service
Rocky Mountain Region
740 Simms Street
PO Box 25127
Lakewood, CO 80225
phone: (303) 275-5350
TTY: (303) 275-5367

Arapaho-Roosevelt National Forests
Pawnee National Grassland
phone: (970) 498-2770

Grand Mesa-Umcompahgre
Gunnison National Forests
phone: (970) 874-7641

Pike-San Isabel National Forests
Commanche & Cimarron National
Grasslands
phone: (719) 545-8737

San Juan-Rio Grande National Forest
phone: (719) 852-5941

White River National Forest
phone: (970) 945-2521

Bureau of Land Management
Colorado State Office
2850 Youngfield St.
Lakewood, Co. 80215-7093
phone: (303) 239-3600
fax: (303) 239-3933
Tdd: (303) 239-3635
Email: msowa@co.blm.gov
Office Hours: 7:45 a.m. - 4:15 p.m.

National Parks

Mesa Verde National Park, CO 81330
phone: (303) 529-4465

Rocky Mountain National Park
phone: (303) 586-2371

Associations, Publications, etc.

Colorado Dude & Guest Ranch Assoc.
PO Box 300
Tabernash, CO 80478
phone: (970) 887-3128
directory: (970) 887-9248
fax: (970) 887-2456

The Dude Ranchers' Association
PO Box F-471
LaPorte, CO 80535
phone: (970) 223-8440
fax: (970) 223-0201

The Dude & Guest Ranches of Grand
County
phone: (800) 247-2636
http://dude-ranch.com/body/html

DudeRanches.com
http://www.duderanches.com

License and Report Requirements

- State requires licensing of Outdoor Professionals.
- State requires an "Inter-Office Copy of Contract with Client" be submitted each time a client goes with an Outfitter. Colorado Agencies of Outfitters Registry sends this copy to client to fill out and return to their agency.

Beaver Meadows Resort Ranch

Don and Linda Weixelman
100 Marmot Dr. #1 • PO Box 178 • Red Feather Lakes, CO 80545
phone: (800) 462-5870 • (970) 881-2450 • fax: (970) 881-2643
email: bmrr@verinet.com

Beaver Meadows is a full-service destination resort with a relaxed, family-oriented atmosphere in a breathtaking setting. Located on the North Fork of the Cache La Poudre River, our ranch occupies 320 acres of mountain meadows, willow creeks, lodgepole forests and quaking aspen groves.

Open year-round, we've provided group and vacation services for more than 20 years. Our staff is committed to providing individual attention and quality service to every event that we host.

We offer year-round activities. Our 22-mile trail system is used for extensive equine activities, mountain biking, hiking, cross-country skiing and showshoeing. Lessons and guides are available for all these activities. Activities are not included in our nightly rates unless a package is requested.

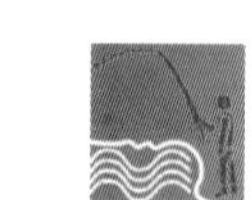

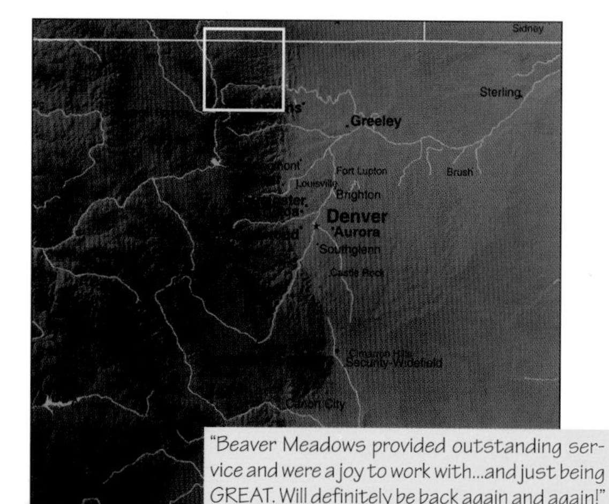

"Beaver Meadows provided outstanding service and were a joy to work with...and just being GREAT. Will definitely be back again and again!"
Denise Noble

Beaver Meadows Resort Ranch

Echo Canyon Guest Ranch

David Hampton

P.O. Box 328 • La Veta, CO 81055

phone: (800) 341-6603 • (719) 742-5524 • fax: (719) 742-5525

email: echo@rmi.net • www.guestecho.com • Lic. #1143

Echo Canyon affords its guests a "unique western adventure." We're proud of our quality riding program for beginners as well as riders who can work cattle. We match our riders with athletic horses that we own and train.

Your "unique western adventure" includes trail rides, roping lessons, cattle work, overnight pack trip, cookout, cowboy entertainment, trout ponds, shooting range, 4 x 4 tours, game area, hot tub, delicious food, quality rooms and cabins.

Our scenery is spectacular. We are located at 8,500 feet elevation beneath West Spanish Peak in Southern Colorado where wildlife abounds.

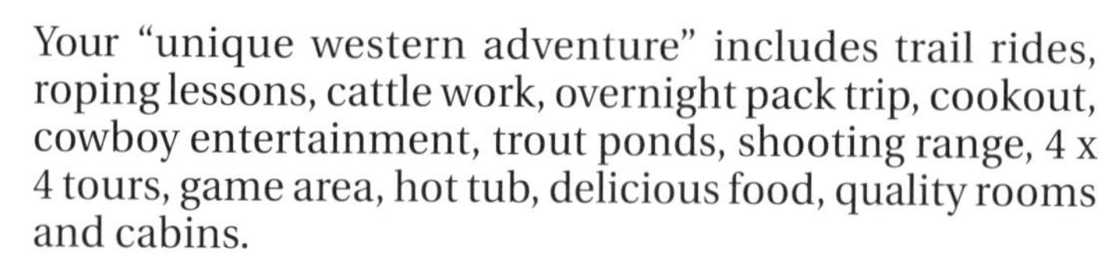

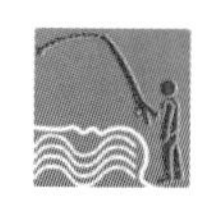

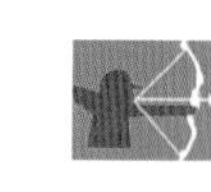

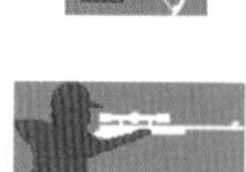

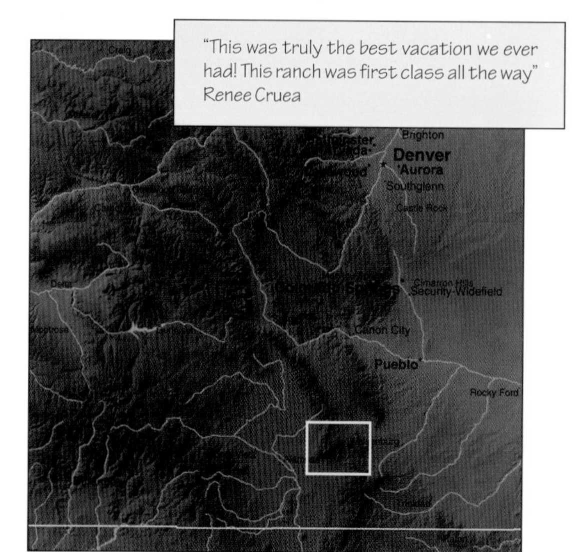

"This was truly the best vacation we ever had! This ranch was first class all the way"
Renee Cruea

ECHO CANYON
E
Colorado
GUEST RANCH

Frazier Outfitting

Sammy Frazier
HC 34, Box 81 • Rye, CO 81069
phone: (719) 676-2964 • license #1738

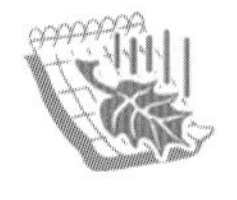
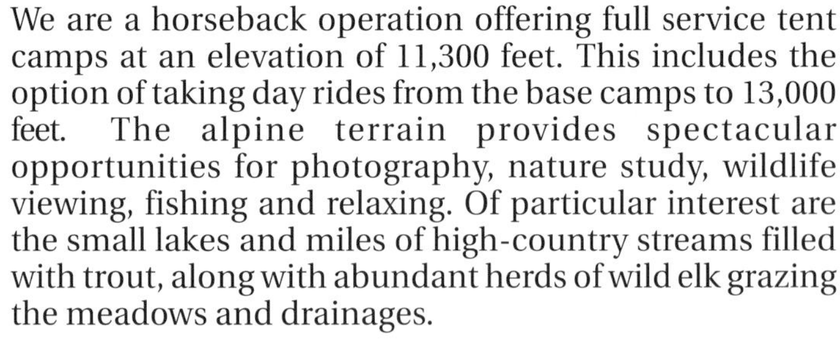

We are a horseback operation offering full service tent camps at an elevation of 11,300 feet. This includes the option of taking day rides from the base camps to 13,000 feet. The alpine terrain provides spectacular opportunities for photography, nature study, wildlife viewing, fishing and relaxing. Of particular interest are the small lakes and miles of high-country streams filled with trout, along with abundant herds of wild elk grazing the meadows and drainages.

We also offer pack-in/out service for campers and backpackers.

Frazier Outfitting operates 40 miles southwest of Creede, Colorado, at the headwaters of the Rio Grande River and surrounded by the Continental Divide. Permitted access includes both the Rio Grande National Forest and the Weminuche Wilderness.

This is a small business specializing in personal, quality, outdoor experience.

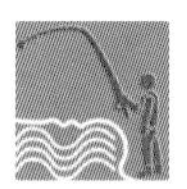

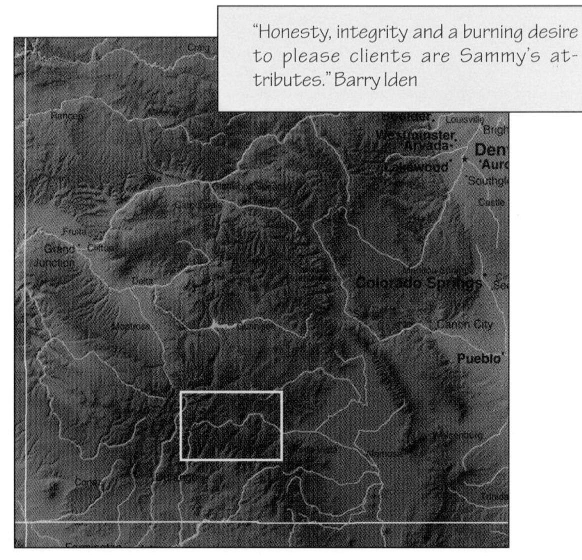

FRAZIER
OUTFITTING

Lakeview Resort and Outfitters

Dan and Michelle Murphy
Box 1000 • Lake City, CO 81235
phone: (800) 456-0170 • (970) 944-2401 • fax: (970) 641-5952 * Lic. # 939

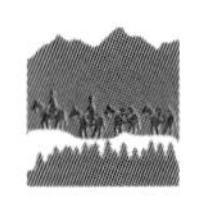

The Lakeview Resort is located on the shores of Lake San Cristobal, in many folks' opinion, the most beautiful lake in Colorado.

We have lodge rooms, spacious suites and quaint cabins with woodburning fireplaces or stoves.

The Lakeview Resort offers a family atmosphere with many adventurous activities. We offer fishing boats, family pontoon boats and new Jeeps for rent. Expert guided fishing is available on the lake.

Exciting horseback activities are also available, with two-hour rides, sunset supper rides, all-day historic or fishing rides and overnight pack trips.

We have the most complete conference center facility in the Lake City area.

"I cannot praise them enough. They went above and beyond the call of duty to make my experience a treasured memory."

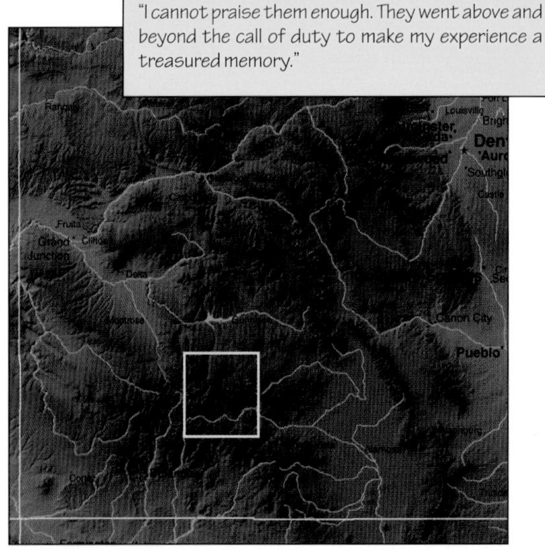

LAKEVIEW
RESORT, INC.
& OUTFITTERS, INC.

San Juan Outfitting

Tom and Cheri Van Soelen
186 County Rd. 228 • Durango, CO 81301
phone: (970) 259-6259 • fax: (970) 259-2652
email: sjo@frontier.net • www.subee.com/sjo/ • Lic. # 997

San Juan Outfitting specializes in classic western horse pack trips. We offer only high-quality trips catering to small groups.

Our spring pack trips take you into the ruins of the ancients (Anasazi Indians) for three to four days. Summer and fall pack trips for fishing, photography and relaxing takes you into the heart of the Weminuche Wilderness to a base camp just below the Continental Divide at an elevation of 10,300 feet. Our high country lake trip travels portions of the Divide while fishing some of the high lakes.

The ultimate adventure is our Continental Divide ride. We travel approximately 100 miles of the Divide at an average elevation of 12,500 feet.

"The whole thing from food, to horses, to living quarters, to friendliness, to 'you name it' was perfect!" Spencer McLean

Schmittel Packing & Outfitting

David and Verna Schmittel
15206 Hwy. 285 • Saguache, CO 81149
phone: (719) 655-2722 • Lic. # 344

Schmittel Packing and Outfitting has provided exciting, high-quality pack trips for 30 years in the wilderness and non-wilderness areas of the San Juan/Rio Grande and Gunnison National Forests, located in "America's Switzerland" of southwestern Colorado.

The area provides a spectacular opportunity to enjoy excellent trail horses, abundant and varied wildlife, unique riding experiences and excellent meals. Dave and Verna have hosted guests with varied riding abilities from every state and many foreign countries.

No one has to be a seasoned rider to enjoy the gorgeous scenery and good company. Each trip is different, allowing guests to visit this magnificent region again and again.

Member of Colorado Outfitters Association, Rio Grande Chapter of the Colorado Outfitters Association, and People for the West.

"It is a privilege and a pleasure to horsepack with Dave and Verna Schmittel, and the finest string of pack horses I have ever experienced" Ronald F. Cox

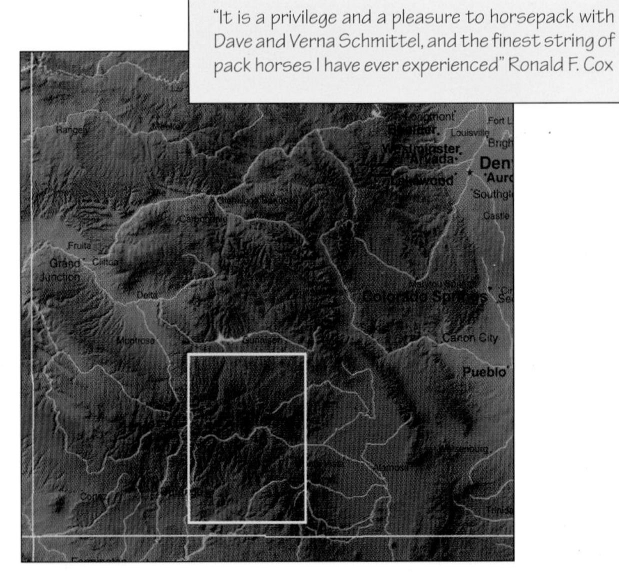

Skyline Guest Ranch

Sheila and Mike Farny

PO Box 67 • Telluride, CO 81435

phone: (888) 754-1126 • (970) 728-3757 • fax: (970) 728-6728

email: skyline-ranch@toski.com

A warm western welcome awaits you, your family and friends at Skyline Guest Ranch. We are committed to sharing with you a special spirit we call "Mountain Joy." Camaraderie flourishes, adventures are shared and there is time for special moments in surroundings of unsurpassed beauty and peace.

You may choose to spend your holiday in one of our ten lodge rooms, each with private bath, or in one of our housekeeping cabins which sleep from two to six people. Skyline is located three miles from the Telluride ski area where you will find some of the finest, uncrowded slopes in the West.

In winter, we offer sleigh rides, cross country skiing, and fine dining. In summer you will enjoy riding one of our horses, fishing in our three trout-filled lakes or riding a mountain bike to an abandoned ghost town.

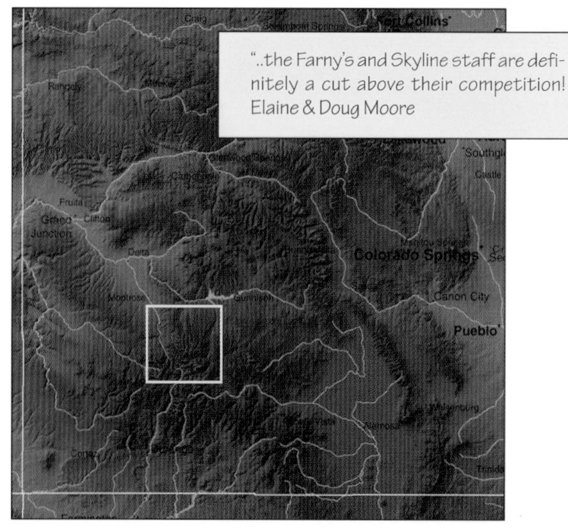

"..the Farny's and Skyline staff are definitely a cut above their competition!
Elaine & Doug Moore

SKYLINE
GUEST RANCH

Picked-By-You Professionals in

Idaho

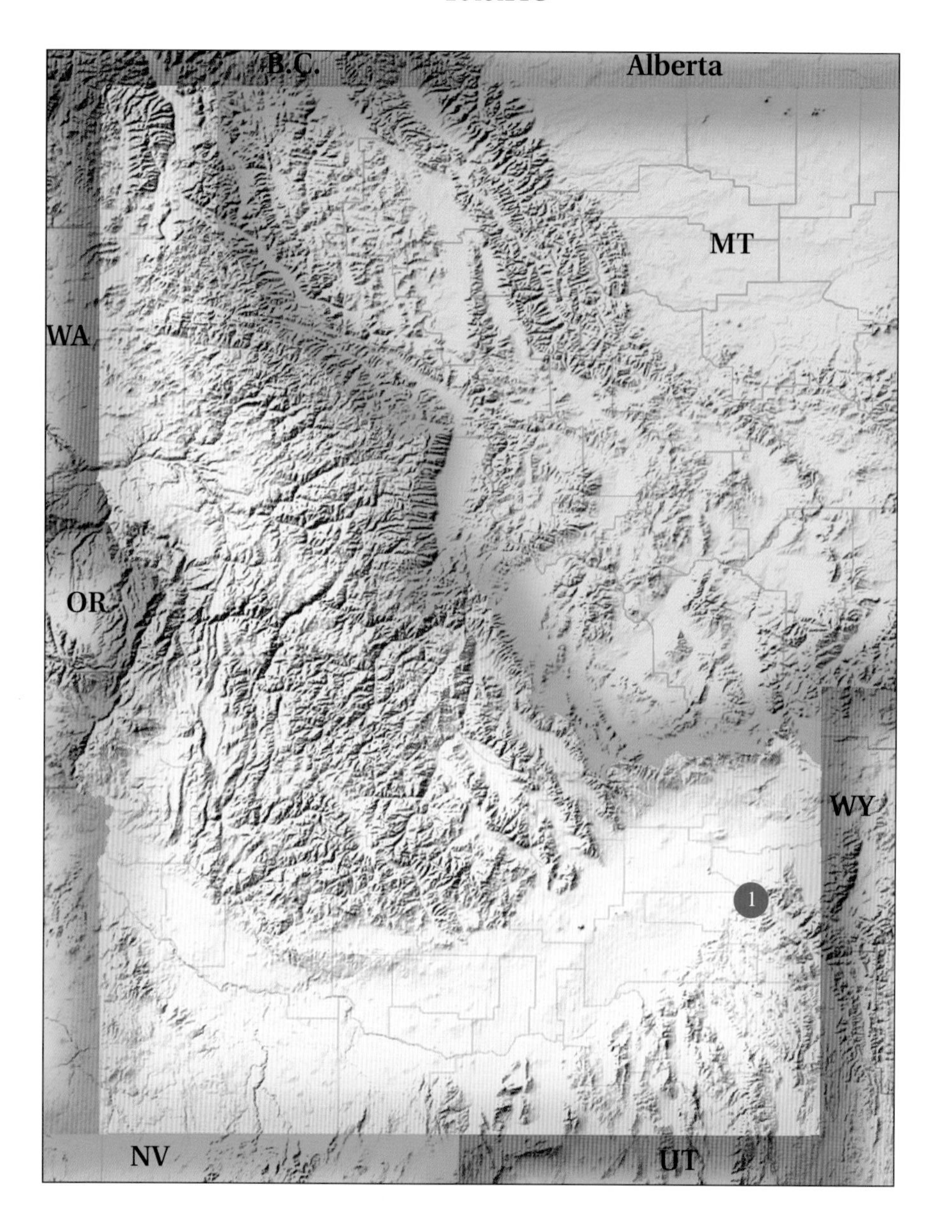

Outdoor Professionals

1 Granite Creek Guest Ranch

Useful information for the state of

Idaho

State and Federal Agencies

Outfitter & Guides Licensing Board
1365 N. Orchard, Room 172
Boise, ID 83706
phone: (208) 327-7380
fax: (208) 327-7382

Idaho Fish & Game Dept.
600 South Walnut
Boise, ID 83707
phone: (208) 334-3700

Forest Service
Northern Region
Federal Bldg.
PO Box 7669
Missoula, MT 59807-7669
phone: (406) 329-3616
TTY: (406) 329-3510

Clearwater National Forest
phone: (208) 476-4541

Idaho Panhandle, Coeur d'Alene-
Kaniksu-St. Joe National Forests
phone / TTY: (208) 765-7223

Nez Perce National Forest
phone: (208) 983-1950

Bureau of Land Management
Idaho State Office
1387 S. Vinnell Way
Boise, ID 83709-1657
phone: (208) 373-3896
or (208) 373-plus ext.
fax: (208) 373-3899

Office Hours 7:45 a.m. - 4:15 p.m.

Associations, Publications, etc.

Idaho Guest and Dude Ranch Assoc.
HC 72
Cascade, ID 83611
phone: (208) 382-4336
message phone: (208) 382-3217

DudeRanches.com
http://www.duderanches.com

Idaho Outfitters & Guides Association
PO Box 95
Boise, ID 83701
phone: (208) 342-1438

License and Report Requirements

- State requires licensing of Outdoor Professionals.
- State requires that every Outfitter be it bird, fish, big game, river rafting, trail riding or packing file a "Use Report" annually.
- Currently, no requirements for Guest/Dude Ranches.

Granite Creek Guest Ranch

Carl and Nessie Zitlau

P.O. Box 340 • Ririe, ID 83443

phone: (208) 538-7140 • fax: (208) 538-7876

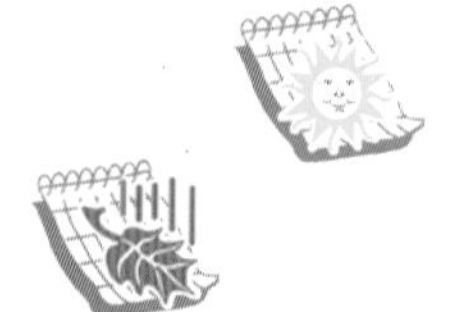

Granite Creek Guest Ranch is one of the most scenic cattle ranches in the West. Nestled against the border of Caribou National Forest, it is comprised of about 2,600 acres of mountainous timber and range land, and 400 acres of farm land. The Zitlau family has raised cattle here since the early 1900s.

It is a "real" working cattle ranch with just the right touch of civilization. Families, couples and singles of all ages and horse skills have a terrific time.

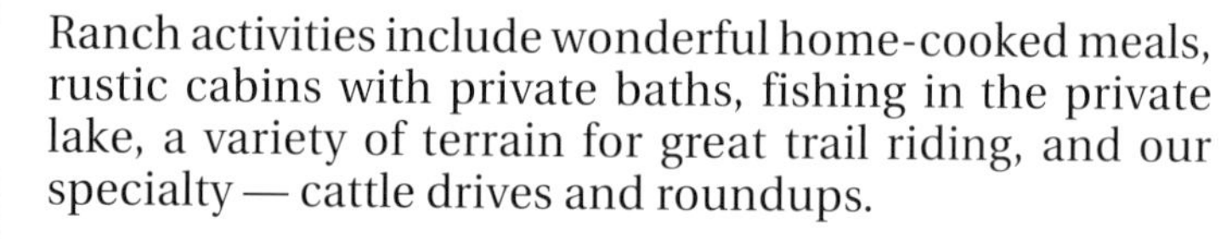

Ranch activities include wonderful home-cooked meals, rustic cabins with private baths, fishing in the private lake, a variety of terrain for great trail riding, and our specialty — cattle drives and roundups.

"My 16 year old son enjoyed this more than anything else we've ever done, and we've done a lot" Kathy Crowley

GRANITE CREEK
GUEST RANCH

Picked-By-You Professionals in

Montana

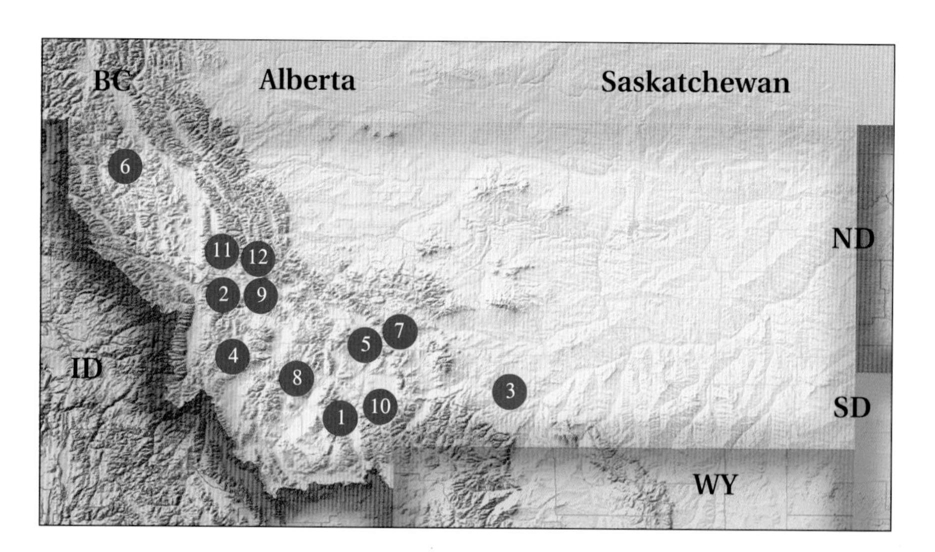

Outdoor Professionals

1. Broken Arrow Lodge
2. Cheff Guest Ranch
3. Double Spear Ranch
4. Esper's Under Wild Skies Lodge & Outfitters
5. EW Watson & Sons Outfitting
6. Hargrave Cattle & Guest Ranch
7. Hidden Hollow Hideaway
8. Iron Wheel Ranch
9. Monture Face Outfitters
10. Nine Quarter Circle Ranch
11. Rich Ranch, LLC
12. White Tail Ranch/WTR Outfitters, Inc.

Useful information for the state of

Montana

State and Federal Agencies

Montana Board of Outfitters
Dept. of Commerce
Arcade Building - 111 North Jackson
Helena, MT 59620-0407
phone: (406) 444-3738

Montana Dept. of Fish, Wildlife & Parks
1420 East 6th
Helena, MT 59620
phone: (406) 444-2535

Forest Service
Northern Region
Federal Building
PO Box 7669
Missoula, MT 59807-7669
phone: (406) 329-3616
TTY: (406) 329-3510

Bitterroot National Forest
phone: (406) 363-7117

Custer National Forest
phone / TTY: (406) 657-6361

Flathead National Forest
phone: (406) 755-5401

Gallatin National Forest
phone / TTY: (406) 587-6920

Helena National Forest
phone: (406) 449-5201

Kootenai National Forest
phone: (406) 293-6211

Lewis & Clark National Forest
phone: (406) 791-7700

Lolo National Forest
phone: (406) 329-3750

Bureau of Land Management
Montana State Office
Granite Tower
222 North 32nd Street
P.O. Box 36800
Billings, Montana 59107-6800
phone: (406) 255-2885
fax: (406) 255-2762
Email - mtinfo@mt.blm.gov
Office Hours: 8:00 a.m. - 4:30 p.m.

National Parks

Glacier National Park
phone: (406) 888-5441

Associations, Publications, etc.

Montana Big Sky Ranch Association
1627 West Main Street, Suite 434
Bozeman, MT 59715

DudeRanches.com
http://www.duderanches.com

License and Report Requirements

- State requires licensing of Outdoor Professionals.

- State requires an "Annual Client Report Log" for all Hunting and Fishing Outfitters.
- State does not regulate River Guides.

- Guest/Dude Ranches need to get an Outfitter license only if they take guest to fish or hunt on land that they do not own.

Broken Arrow Lodge

Erwin and Sherry Clark
2200 Upper Ruby Rd., Box 177 • Alder, MT 59710
phone: (800) 775-2928 • phone/fax: (406) 842-5437
www.recworld.com/state/mt/hunt/broken/broken.html

Broken Arrow Lodge is located in the Snowcrest Mountain Range in Southwest Montana's Ruby Valley. The Ruby River flows through the property and is only a moment's walk away.

Broken Arrow Lodge is a modern facility known for friendly, personalized service in a homey atmosphere. We supply lodging, meals (served at your convenience), and year-round recreation. Activities include fishing, hunting, family vacations, horseback riding, wildlife viewing, winter sports, lodge activities, and more.

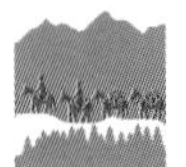

Five rooms are available with space for one to four, or a family-size room with space for up to eight. Rooms are clean and comfortable with your choice of single or double beds. Large front deck provides a great area to relax and view the breathtaking scenery, abundant wildlife, and beautiful wildflowers.

Airport shuttle service is available as well as equipment rental, fax, and satellite TV.

"The hospitality was wonderful...we felt like we were visiting friends" Mary Ann McGuire

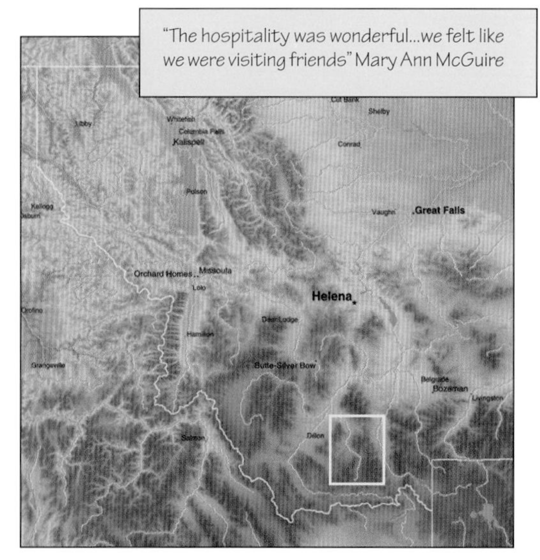

BroKen Arrow
ULTIMATE OUTDOOR EXPERIENCE
MONTANA

Cheff Guest Ranch

Mick and Karen Cheff
4274 Eagle Pass Rd. • Charlo, MT 59824
phone: (406) 644-2557

A world of wondrous natural beauty and superb outdoor recreation awaits you at the Cheff Guest Ranch.

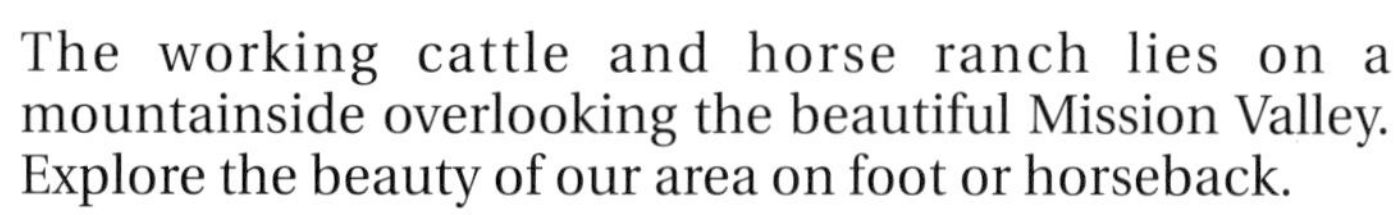

The working cattle and horse ranch lies on a mountainside overlooking the beautiful Mission Valley. Explore the beauty of our area on foot or horseback.

Flathead Lake, the National Bison Range, and Glacier National Park are just a few of the scenic and historic attractions located nearby. Fishing and scenic pack trips in the Bob Marshall and Mission Mountain Wilderness begin in early July. They are a once-in-a-lifetime experience, yet many take the trips repeatedly.

We are one of Montana's oldest outfitting families with more than 65 years of experience. Our experience and desire to please you combine for a memorable, and we hope, successful trip.

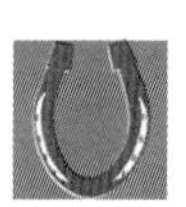

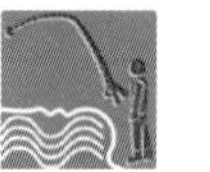

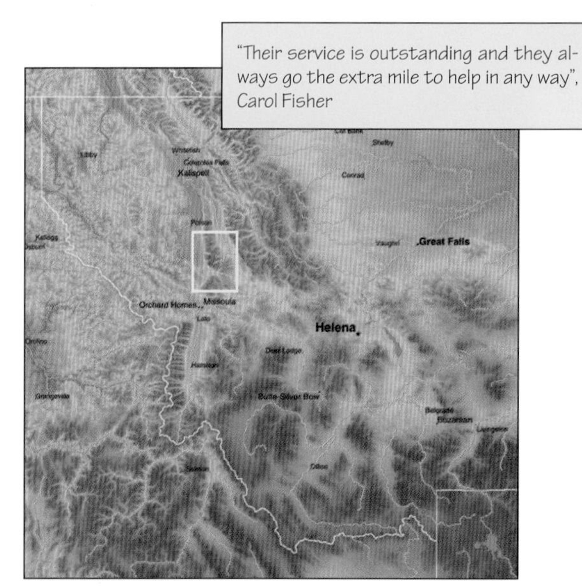

"Their service is outstanding and they always go the extra mile to help in any way",
Carol Fisher

CHEFF GUEST
RANCH
Outfitters and Guides

Double Spear Ranch

Tony and Donna Blackmore
P.O. Box 43 • Pryor, MT 59066
phone: (406) 259-8291 • fax: (406) 245-7673
email: DoubleSpearRanch@juno.com

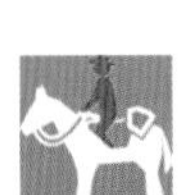

We invite adventurous city slickers to join us and live the real West for a week or two on our working cattle ranch 35 miles south of Billings, Montana, on the Crow Indian Reservation. Although a relative of old Sitting Bull, your cowboss Tony Blackmore will remind you more of John Wayne. Ride the range and the mountains, work cattle, learn colt-breaking and horsemanship techniques, and enjoy cowboy cookouts. You will ride quarter horses and meet rare (hypoallergenic!) American curly horses — buffalo — and lots more livestock and pets. This isn't a fancy upscale vacation; you will join in and experience real ranch life. Expect a little dust, a little sweat, and tons of laughter.

We include your own horse, unlimited riding, ranch meals cowboy-style, airport pickup and delivery. Bedrooms for adults in ranch house or bunkhouse (shared facilities).

Special 2-week western youth camps for guests 13-19.

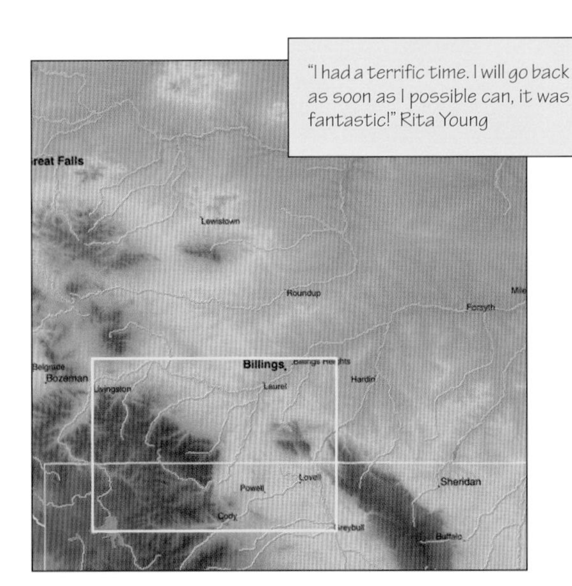

DOUBLE SPEAR RANCH

Esper's Under Wild Skies Lodge & Outfitters

Vaughn and Judy Esper
P.O. Box 849 • Philipsburg, MT 59858
phone: (406) 859-3000 • fax: (406) 859-3161

Under Wild Skies Lodge and Outfitters is located in the Deerlodge National Forest at the boundary of the Anaconda Pintler Wilderness.

Our guest ranch offers something for everyone. For the fisherman we have two lakes on the ranch. The Middle Fork of Rock Creek traverses through the property and offers four species of trout. Take a scenic wilderness horseback ride for a day or an overnight pack trip into the majestic Pintler Mountains. Or, just relax in the casual elegance of the lodge.

At Under Wild Skies we take pride in our facilities, services, and the meticulous attention we pay to every detail of your stay. You come as a guest and leave as a friend.

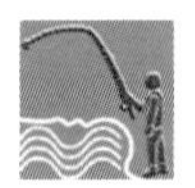

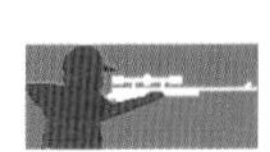

"Wonderful experience, beautiful location, great cooking, 'unmatched'!" Mr. & Mrs. Ronald Vachon

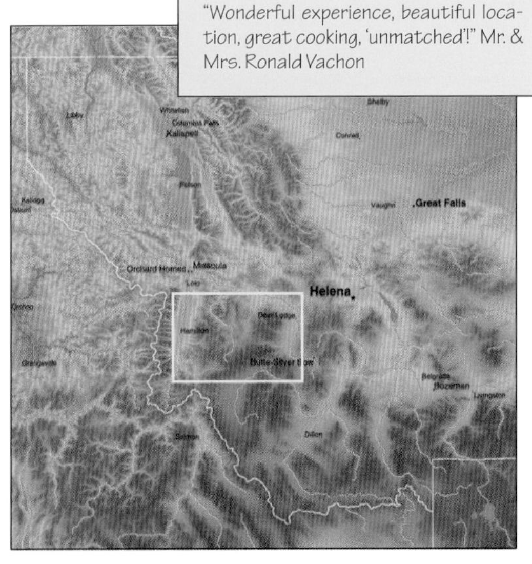

ESPER'S
LODGE & OUTFITTERS
UNDER WILD SKIES
MONTANA

EW Watson & Sons Outfitting

Ed and Wanda Watson
7837 U.S. Hwy. 287 • Townsend, MT 59644
phone: (800) 654-2845 • fax: (406) 266-4498

WHERE THE ONLY THING BETTER THAN THE SCENERY IS THE SERVICE

E W Watson & Sons Outfitting is dedicated to providing top hands, quality horses, and an educational and affordable vacation.

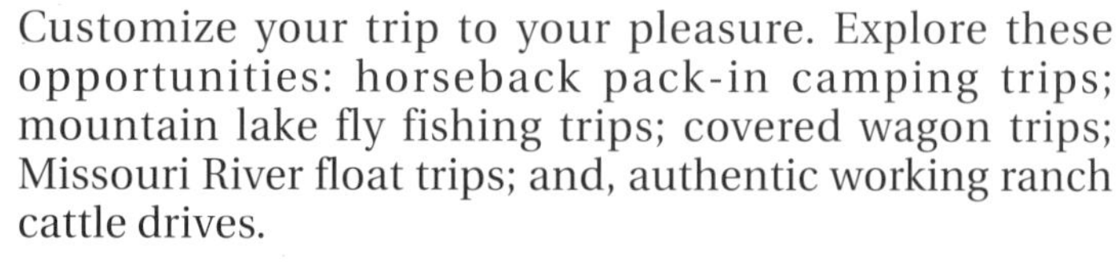

Customize your trip to your pleasure. Explore these opportunities: horseback pack-in camping trips; mountain lake fly fishing trips; covered wagon trips; Missouri River float trips; and, authentic working ranch cattle drives.

Combine local historical points of interest with ranch home lodging. Visit Elkhorn Ghost Town, tour Lewis and Clark Caverns, ride the tour train at Helena and learn about the colorful gold rush days.

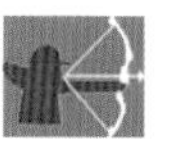

EW

Hargrave Cattle and Guest Ranch

Leo and Ellen Hargrave
300 Thompson River Rd. • Marion, MT 59925
phone: (406) 858-2284 • fax: (406) 858-2444
email: hargrave@digisys.net • www.hargraveranch.com

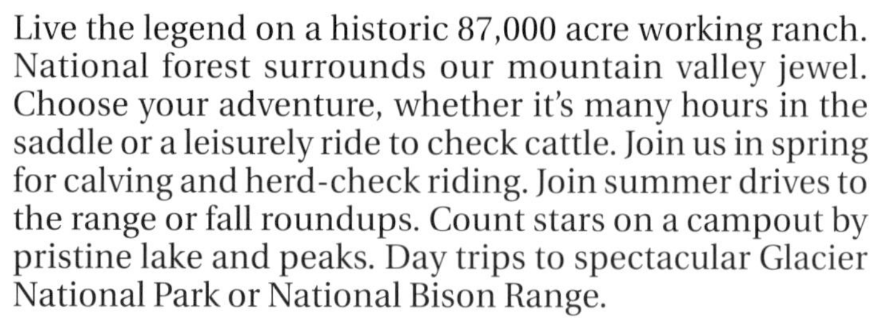

Live the legend on a historic 87,000 acre working ranch. National forest surrounds our mountain valley jewel. Choose your adventure, whether it's many hours in the saddle or a leisurely ride to check cattle. Join us in spring for calving and herd-check riding. Join summer drives to the range or fall roundups. Count stars on a campout by pristine lake and peaks. Day trips to spectacular Glacier National Park or National Bison Range.

Skeet and target shooting, archery, lake canoeing, cowboy campfire sing-a-longs, pool games in the horse barn museum, and private meadow fishing. On-site massage therapist arranged.

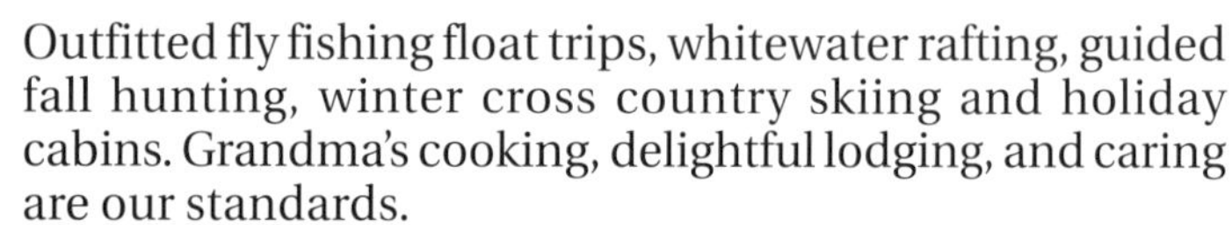

Outfitted fly fishing float trips, whitewater rafting, guided fall hunting, winter cross country skiing and holiday cabins. Grandma's cooking, delightful lodging, and caring are our standards.

We were Outfitters on the Great Montana Cattle Drive and are committed to sharing that Western Spirit with our guests.

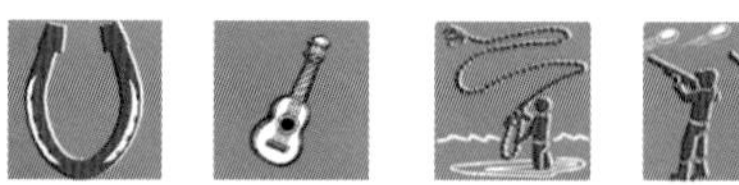

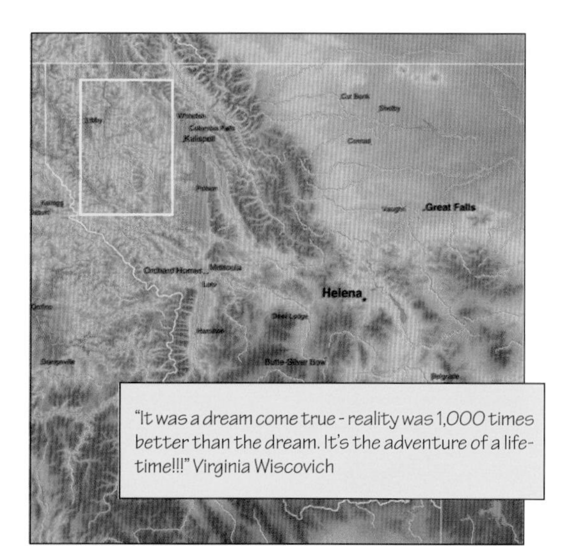

HARGRAVE CATTLE & GUEST RANCH
HARGRAVE
THOMPSON RIVER VALLEY
MARION. MONTANA
HARGRAVE
CHARDLAIS RANCH

Hidden Hollow Hideaway

Kelly and Jill Flynn
P.O. Box 233 • Townsend, MT 59644
phone: (406) 266-3322

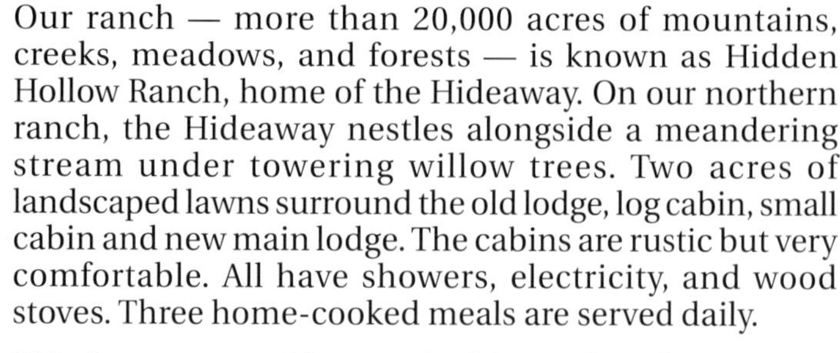

Our ranch — more than 20,000 acres of mountains, creeks, meadows, and forests — is known as Hidden Hollow Ranch, home of the Hideaway. On our northern ranch, the Hideaway nestles alongside a meandering stream under towering willow trees. Two acres of landscaped lawns surround the old lodge, log cabin, small cabin and new main lodge. The cabins are rustic but very comfortable. All have showers, electricity, and wood stoves. Three home-cooked meals are served daily.

Ride horses on a ridge overlooking miles of mountains and meadows. Pan for gold alongside a rushing mountain stream. Take an "off the beaten path" four-wheel-drive tour. Sit around a campfire or enjoy a barbecue. Hike through wildflower-blazing meadows. Fish at a nearby creek or one of our ranch ponds. Pitch in on some of the ranch work, or just sit back and enjoy the peace and solitude.

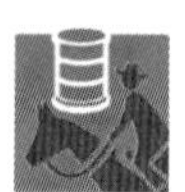
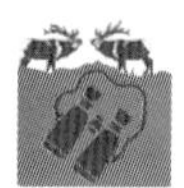
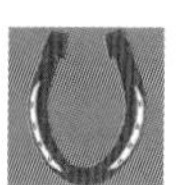

"Kelly Flynn makes it all happen!" Marion Dial

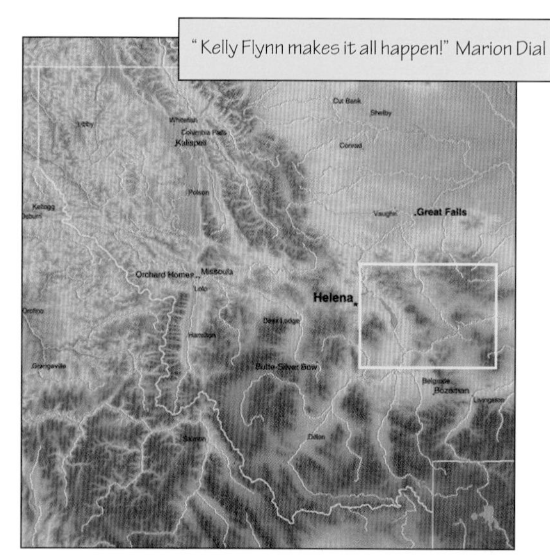

Hidden Hollow Hideaway

Iron Wheel Ranch

John & Sherry Cargill
40 Cedar Hills Road •Whitehall, MT 59759
phone/fax: (406) 494-2960 • mobile: (406) 491-2960

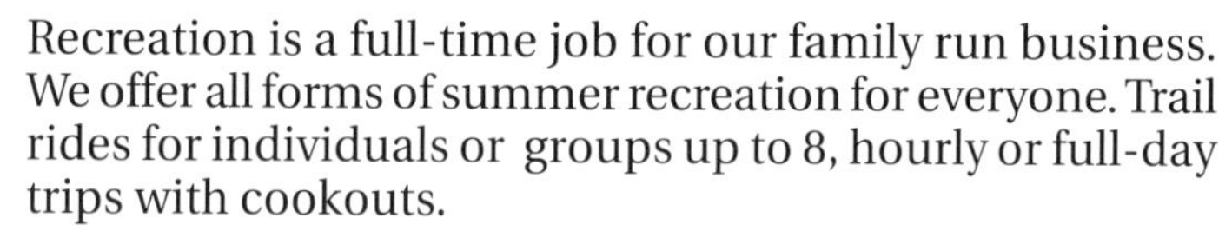

Recreation is a full-time job for our family run business. We offer all forms of summer recreation for everyone. Trail rides for individuals or groups up to 8, hourly or full-day trips with cookouts.

Our youth camps for children 7-15 specialize in teaching children how to handle, saddle, care and ride their horse. Child gets their "own" horse for the entire week. We camp out and children put up their own tent and learn other camping responsibilities.

Blue ribbon rivers are fun for everyone, with many river and fishing option available as well as our private pond.

Our Bed & Breakfast lodge offers many combination vacations including float trips, trail rides, seasonal varmint and big game hunts. On site we have horseshoe pits, volleyball net, BBQ's, fire rings and a creek. We are located near the Continental Divide and easily accessible.

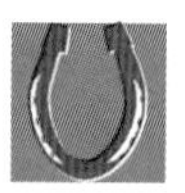

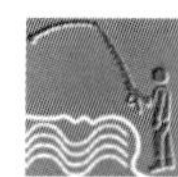

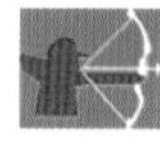

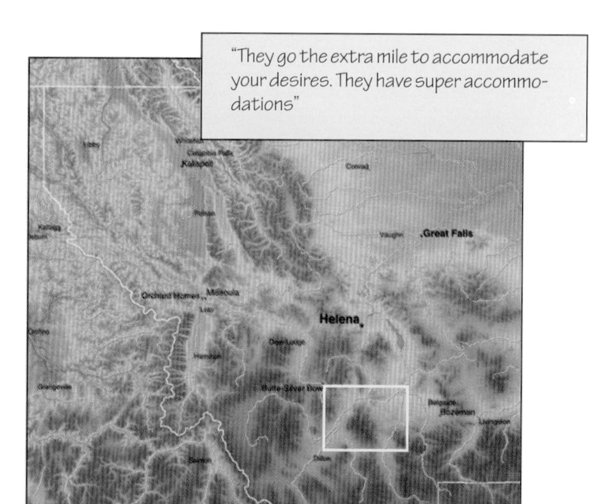

IRON WHEEL RANCH
Youth Camps
Trail Rides
Big Game Hunts
Float Trips
Bed-n-Breakfast

Monture Face Outfitters

Tom Ide and Valerie Call
Box 27 • Greenough, MT 59836
phone: (888) 420-5768 • phone/fax: (406) 244-5763

Monture Face Outfitters is your host to Montana's "Bob Marshall Wilderness," nearly 1.5 million acres of pristine beauty in the heart of the Rocky Mountains.

Travel into and through the wilderness on top-notch horses and mules. Pack trips are flexible, from three days to a high adventure eight-day roving experience. Abundant wildflowers and wildlife provide photo opportunities at every bend of the trail, and the trout fishing is outstanding. Only the highest quality ingredients are used in the gourmet wilderness kitchen. Build campfires at night and immerse yourself in a sea of stars.

Owner Tom Ide, son Tim, and Valerie Call have one ultimate goal; to share the magic of wilderness. Knowledgeable, experienced and quality -oriented.

"Everything was well planned and executed. The food was fabulous....I would enthusiastically recommend this outfitter to anyone!" Bill Bailey

MONTURE
FACE
Outfitters
Monture Face Outfitters

Nine Quarter Guest Ranch

Kim and Kelly Kelsey

5000 Taylor Fork Rd. • Gallatin Gateway, MT 59730

phone: (406) 995-4276 (ranch) • (406) 586-4972 (home)

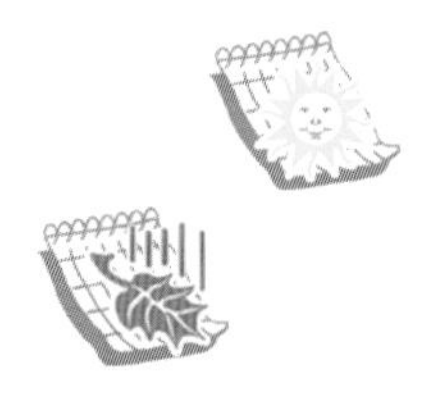

From a secluded valley overlooking Yellowstone National Park, the Kelsey's have been hosting guests for more than 50 years. The ranch-raised herd of 120 Appaloosa horses will take you through pine forests, over mountain streams and across wildflower-strewn meadows.

Fun at the Nine Quarter includes children with a kiddie wrangler and a ranch babysitter. Among the weekly activities are square dances, hay rides, wildlife lectures and softball games. Other pastimes include hiking, photography or just relaxing on your porch to the cry of a coyote.

The Taylor Fork, a fine trout stream, flows through the ranch. With a trout pond and ranch guide at your side, you will soon be hooked on fishing these famous headwaters of the Missouri River.

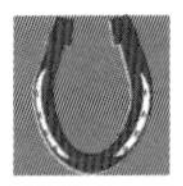

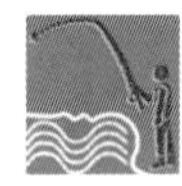

"Our vacation in Montana has become a family tradition, to go back to 'the ranch, our ranch' Nine Quarter Circle Ranch, Gallatin Gateway, Montana"
Mr. & Mrs. Brian Hensley

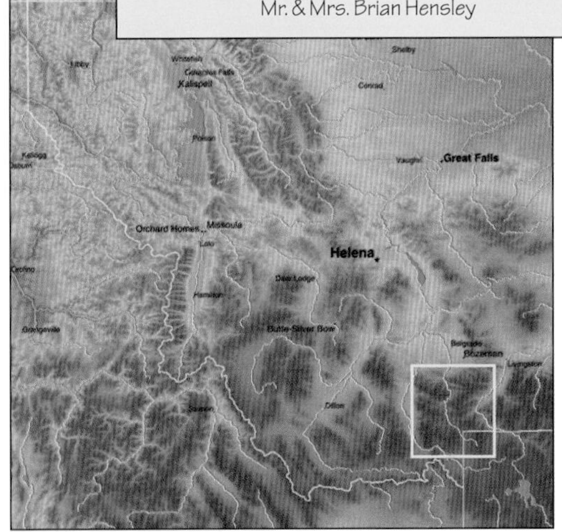

Nine Quarter Guest Ranch

Rich Ranch

Jack and Belinda Rich
P.O. Box 495 • Seeley Lake, MT 59868
phone: (406) 677-2317 • fax: (406) 677-3530
email: richranch@montana.com • www.richranch.com

Summer ranch vacations are available from May through September. While at the ranch you will enjoy quiet country living. The lodge and cabins are nestled in the trees overlooking a large natural meadow with a scenic backdrop of majestic mountains. We are surrounded by more than one million acres of state and national forests.

Horseback riding is the main activity. Each guest is fitted to a saddle and we carefully choose a horse suited to your ability.

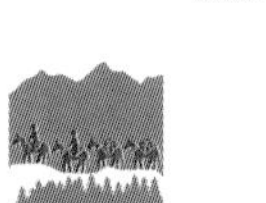

Do you prefer a leisurely morning ride through the meadows and rolling hills, a high adventure trip to the top of the mountain, or maybe some time to work on your horsemanship skills in our outdoor arena? The choice is yours.

Quality fishing for all levels, beginner to expert, is available in nearby lakes, streams and rivers.

"I would have to say our enthusiasm from our trip was infectious enough that other couples and an acquaintant asked to go along on our next trip!" John L. Trudel

Rich Ranch

White Tail Ranch/WTR Outfitters, Inc.

Jack and Karen Hooker
520 Cooper Lake Rd. • Ovando, MT 59854
phone: (888) WTR-5666 • phone/fax: (406) 793-5666
email: wtroutfitters@montana.com • www.recworld.com/wtro

WTR Outfitters has been specializing in summer horse pack trips since 1940.

Join Karen and Jack Hooker on the ride of a lifetime into the Bob Marshall, Great Bear and Scapegoat Wilderness areas.

Hike across fields splashed with the bright blues, reds and yellows of alpine flowers. Ride along the Chinese Wall, a 1,500 foot sheer cliff. Fish for native cutthroat trout in remote, gin-clear streams. Photograph deer, elk, and perhaps even a grizzly.

Trips can be customized to suit your wishes or you can join one of our scheduled trips.

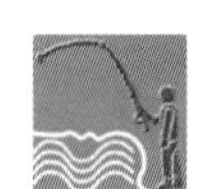

"In my opinion the Hookers' are the very top of their profession!" John & Rose Marie McGoldrick

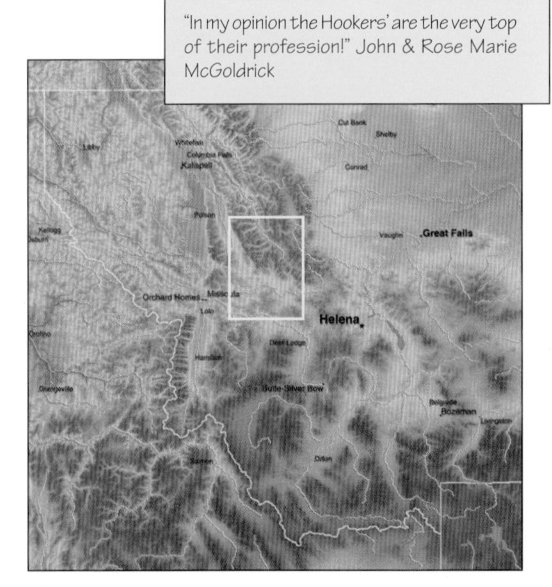

WTR
Outfitters

Picked-By-You Professionals in

New Mexico

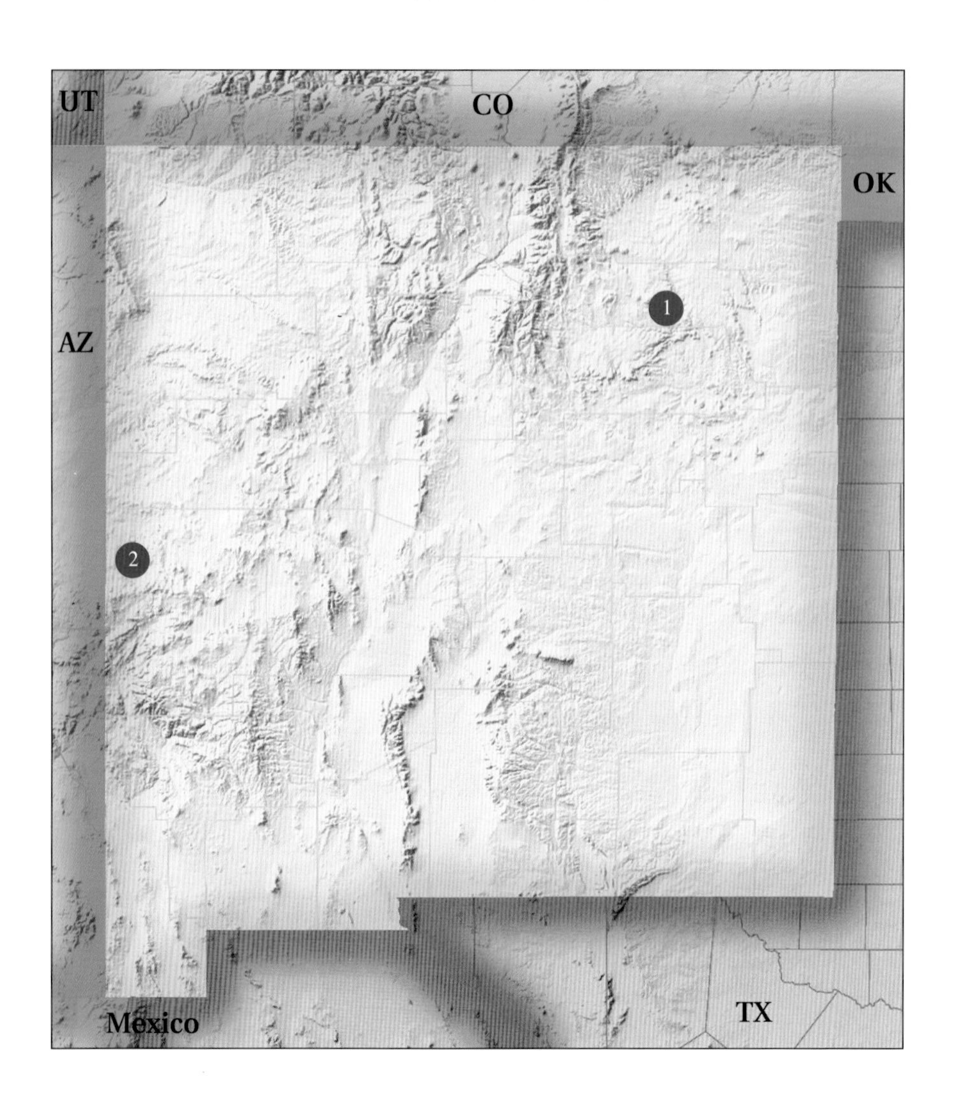

Outdoor Professionals

1. Hartley Guest Ranch
2. Maynard Ranch

Useful information for the state of

New Mexico

State and Federal Agencies

New Mexico Game & Fish Dept.
Villagra Building
Santa Fe, NM 87503
phone: (505) 827-7975

Forest Service
Southwestern Region
Federal Building
517 Gold Avenue SW
Albuquerque, NM 87102
phone: (505) 842-3300
TTY: (505) 842-3898

Carson National Forest
phone: (505) 758-6200

Cibola National Forest
phone / TTY: (505) 761-4650

Gila National Forest
phone: (505) 388-8201

Lincoln National Forest
phone: (505) 434-7200

Santa Fe National Forest
phone: (505) 438-7840

Bureau of Land Management
New Mexico State Office
1474 Rodeo Road
Santa Fe, NM 87505

Mailing Address:
P.O. Box 27115
Santa Fe, NM 87502-0115

Information Number: (505) 438-7400
fax: (505) 438-7435
Public Lands Information Center (PLIC):
(505) 438-7542

Office Hours: 7:45 a.m. - 4:30 p.m.

National Parks

Carlsbad Caverns National Park
3225 National Parks Hwy.
Carlsbad, NM 88220
phone: (505) 785-2232

Associations, Publications, etc.

New Mexico Council of Outfitters & Guides, Inc.
160 Washington SE #75
Albuquerque, NM 87108
phone: (505) 764-2670

DudeRanches.com
http://www.duderanches.com

License and Report Requirements

- State requires that Hunting Outfitters be licensed.
- State requires the filing of an "Annual Report of Outfitters' Clients" for hunting only.
- "Use Permit" required for Fish and River Outfitters using BLM and Forest Service lands. They are not required to file any reports.

Hartley Guest Ranch

Doris and Ray Hartley
HCR 73, Box 55 • Roy, NM 87743
phone: (800) OUR-DUDE (687-3833) • (505) 673-2245 • fax: (505) 673-2216
email: rhart@etsc.net • www.duderanch.org/hartley

So, should you have been a cowboy? The Hartley Family invites you to experience and enjoy our working cattle ranch, nestled in the breathtaking beauty of New Mexico.

Explore 200 miles of trails that circle the rims of redrock canyons and wind through forest of juniper, oak and pine by horseback or ATV. Other ranch activities: cattle working, branding, campouts, fishing, campfires, and hiking. Discover ancient Indian sites, dinosaur bones, and unusual geological formations located on the ranch. Rafting trips and day trips to enchanting Santa Fe and Taos.

Delicious home-cooked meals are served family-style in the dining room or cooked outdoors over an open fire. Transportation from Albuquerque.

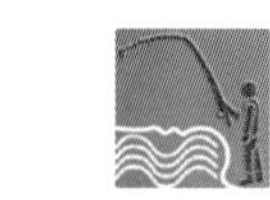

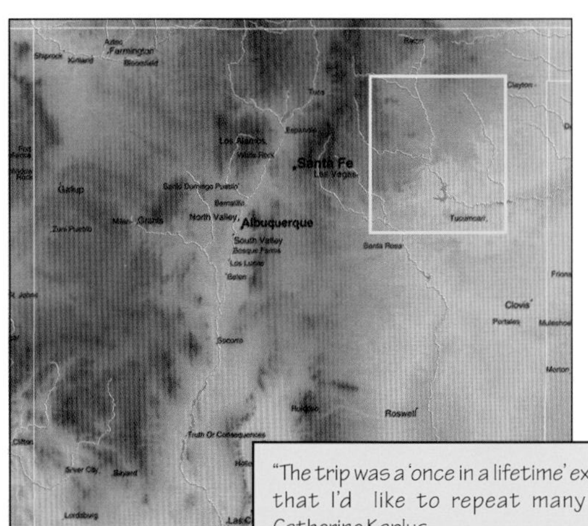

"The trip was a 'once in a lifetime' experience that I'd like to repeat many times!"
Catherine Kaplus

Hartley Guest Ranch

Maynard Ranch

Perry and Brenda Hunsaker • Billy and Nora Maynard
19831 E Warner Rd. • Higley, AZ 85236
phone: (602) 988-9654 • fax: (602) 988-3292

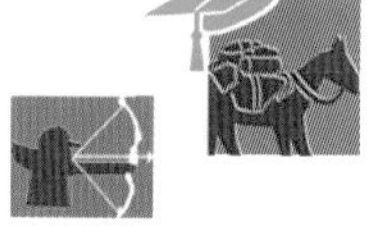

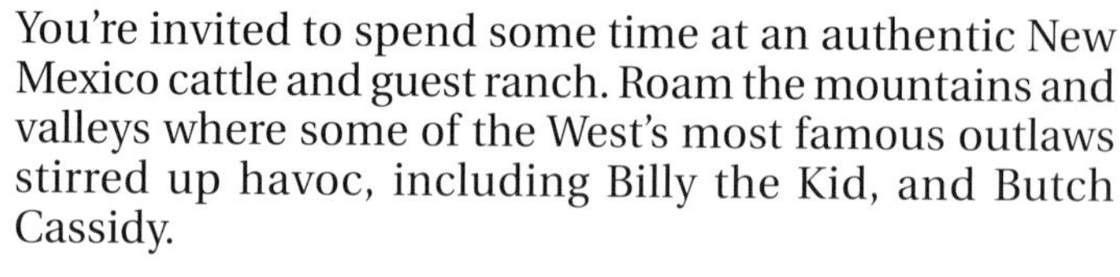

You're invited to spend some time at an authentic New Mexico cattle and guest ranch. Roam the mountains and valleys where some of the West's most famous outlaws stirred up havoc, including Billy the Kid, and Butch Cassidy.

Set on nearly 3,000 acres and surrounded by National Forest with towering ponderosa pines, oak thickets and lush meadows. Enjoy horseback riding through the forests or spend time in the saddle with cowboys working cattle. Play cowboy golf, take a hayride, go fishing or just sit back and unwind.

Daytrips available to nearby Indian ruins and archaeological sites. Spend evenings around the campfire listening to a cowboy poet or enjoy the star-filled sky and Milky Way. We offer a great kids' program.

The ranch has comfortable accommodations perfect for family reunions and corporate retreats. Hearty meals are served family-style.

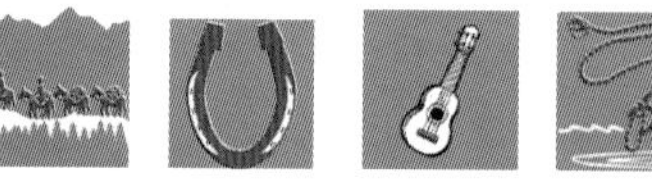

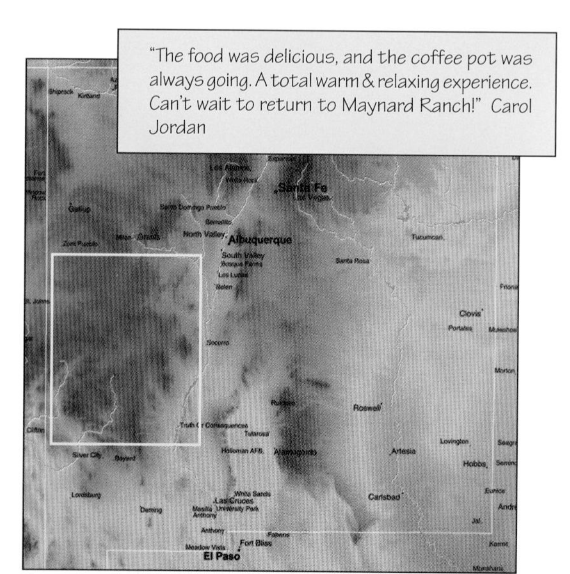

"The food was delicious, and the coffee pot was always going. A total warm & relaxing experience. Can't wait to return to Maynard Ranch!" Carol Jordan

Maynard Ranch

Picked-By-You Professionals in

North Carolina

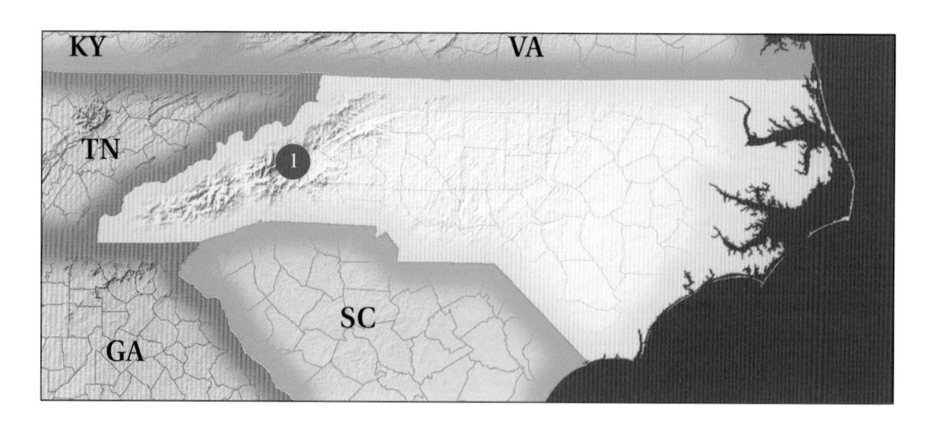

Outdoor Professionals

1 Clear Creek Guest Ranch

Useful information for the state of

North Carolina

State and Federal Agencies

North Carolina Wildlife Resources Commission
Archdale Building
512 N. Salisbury St.
Raleigh, NC 27611
(919) 733-3391

Marine Fisheries
(919) 726-7021

Bureau of Land Management
Eastern States
7450 Boston Boulevard
Springfield, Virginia 22153
Phone: (703) 440-1660
or (703) 440- Plus Extension
Fax: (703) 440-1599

Office Hours: 8:00 a.m. - 4:30 p.m.

Eastern States
Jackson Field Office
411 Briarwood Drive, Suite 404
Jackson, Mississippi 39206
Phone: (601) 977-5400
Fax: (601) 977-5440

Forest Service
Southern Region
1720 Peachtree Road NW
Atlanta, GA 30367
phone: (404) 347-4177
TTY: (404) 347-4278

Croatan-Nantahala-Pisgah-Uwharrie National Forests
phone: (704) 257-4200

Associations, Publications, etc.

DudeRanches.com
http://www.duderanches.com

American Fisheries Society
Box 7617
NC State University
Raleigh, NC 27695
phone: (919) 515-2631

Trout Unlimited North Carolina
135 Tacoma Circle
Asheville, NC 28801-1625
phone: (704) 684-5178
fax: (704) 687-1689

National Hunters Association, Inc.
PO Box 820
Knightdale, NC 27545
phone: (919) 365-7157

Professional Bowhunters Society
PO Box 246
Terrell, NC 28682
phone/fax: (704) 664-2534

Carolina Bird Club, Inc.
PO Box 29555
Raleigh, NC 27626-0555

North Carolina Bass Chapter Federation
1105 Misty Wood Lane
Harrisburg, NC 28075
phone: (704) 785-9108

License and Report Requirements

- State does not license or register Outfitters, Guides, Captains or Lodges.
- State has no report requirements.

Clear Creek Guest Ranch

Rex Frederick
100 Clear Creek Rd., Hwy. 80 South • Burnsville, NC 28714
phone: (800) 651-4510 • (704) 675-4510 • fax: (704) 675-5452
email: clearcreek@mcwalters.net

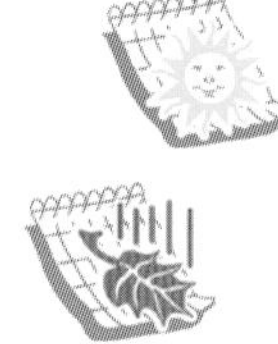

"Best vacation ever! Just like family. ... A piece of heaven. ... Wonderful time, great staff, we'll be back!" These are just a few of the notes we've received from recent guests.

Our goal at CCR is to give you the most relaxing, yet fun-filled vacation possible. Rooms are cozily furnished with lodgepole pine furniture, quilts and all are carpeted, air-conditioned and heated. All buildings have big porches with rockers and offer a magnificent view of the Black Mountains.

Activities include horseback riding, trout fishing, hiking, whitewater rafting and tubing on the South Toe River. A highlight of our week is the Saturday "rodeo."

Come and see for yourself. Clear Creek Ranch brings a bit of the Old West to the mountains of North Carolina. Call for a brochure and more details.

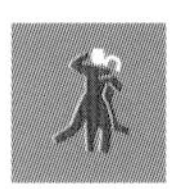

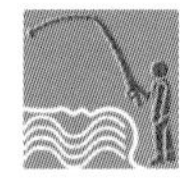

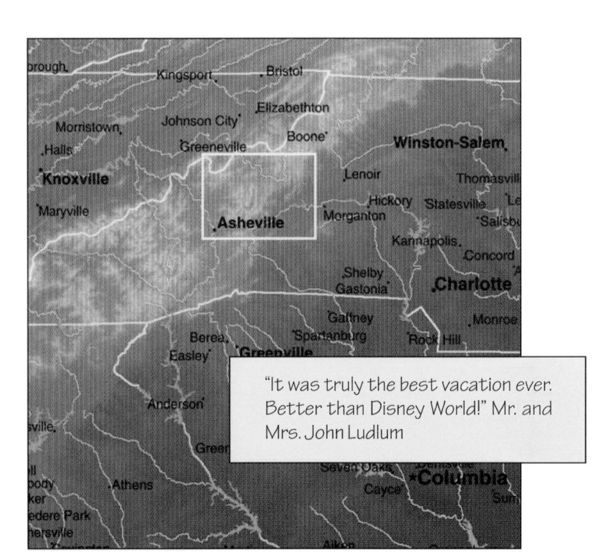

"It was truly the best vacation ever. Better than Disney World!" Mr. and Mrs. John Ludlum

CLEAR CREEK
GUEST RANCH

Picked-By-You Professionals in

North Dakota

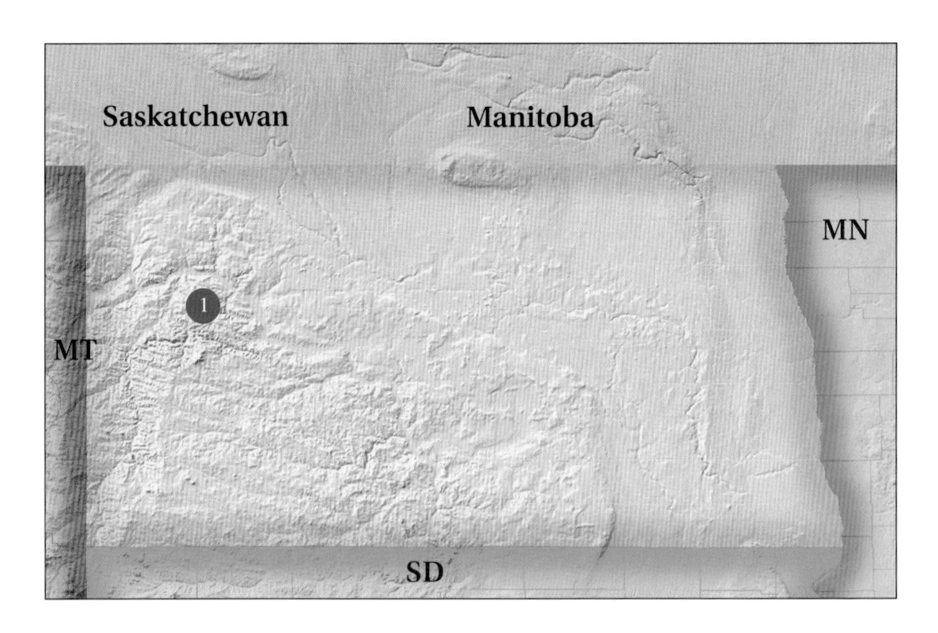

Outdoor Professionals

1. Little Knife Outfitters

Useful information for the state of

North Dakota

State and Federal Agencies

North Dakota Game & Fish Dept.
100 North Bismarck Expressway
Bismarck, ND 58501
(701) 328-6300

Bureau of Land Management
Montana State Office
(serves North & South Dakota also)
222 North 32nd Street
P.O. Box 36800
Billings, Montana 59107-6800

Phone: (406) 255-2885
Fax: (406) 255-2762
Email - mtinfo@mt.blm.gov

Dakotas District Office
2933 Third Avenue West
Dickinson, North Dakota 58601-2619

Phone: (701) 225-9148
Fax: (701) 227-8510
Email - ddomail@mt.blm.gov

Office Hours: 8:00 a.m. - 4:30 p.m.

State Forest Service
307 First Street
Bottineau, ND 58318-1100
phone: (701) 228-5422
fax: (701) 228-5448

National Parks

Theodore Roosevelt National Park
Medora, ND 58645
phone: (701) 623-4466

Associations, Publications, etc.

DudeRanches.com
http://www.duderanches.com

Dakota Outdoors
PO Box 669
Pierre, SD 57501-0669
Phone: (605) 224-7301
Fax: (605) 224-9210

License and Report Requirements

- State does not license or register Outfitters, Guides, or Lodges.
- State has no report requirements.

Little Knife Outfitters

Glendon "Swede" Nelson

RR 1, Box 116 • Stanley, ND 58784
phone: 701-628-2747 • fax: 701-628-3254
email: swede@farside.cc.misu.nodak.edu
http://www.ndcd.org/ndcpd/ndwan/sites/stanley/communtiy/LKO1.html

Would you enjoy an authentic western adventure in the backcountry? Would you enjoy a real cowboy experience replete with horses, campsites, western-style meals, beautiful sunrises, sunsets, wildlife, and nature?

If you answered "yes" to these questions, allow us to introduce ourselves.

We are Swede and Jean Nelson, owners and operators of Little Knife Outfitters. We offer one- to six-day trail ride adventures during June, July and August. We outfit our clients with riding supplies, serve them western-style meals, and guide their excursion through the North Unit of the Theodore Roosevelt National Park in western North Dakota. Our adventure takes us through 24,000 acres of wilderness and secluded backcountry complete with authentic Indian and cowboy cultural artifacts from the days of westward expansion.

Our regularly-scheduled or tailor-made rides depart from the banks of the Little Missouri River.

"I have taken my employees on this ride twice. We all forget the pressure of the business world and communicate to each other about 'real topics' of life. Great time, can't wait to do it again" Roger Gjeiistad-Stanley Equipment, Inc.

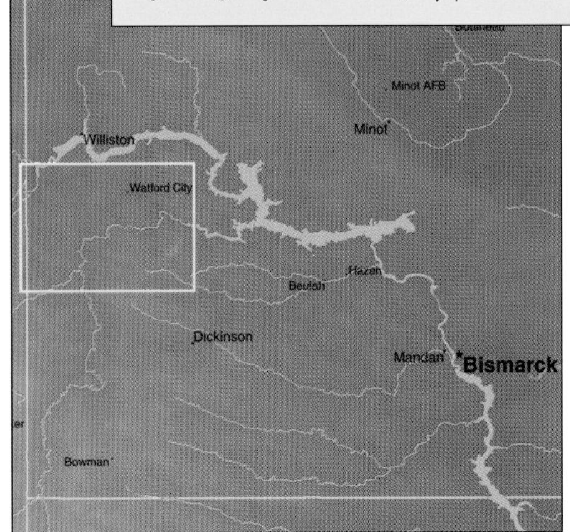

Little Knife Outfitters, Inc.

Picked-By-You Professionals in

Oregon

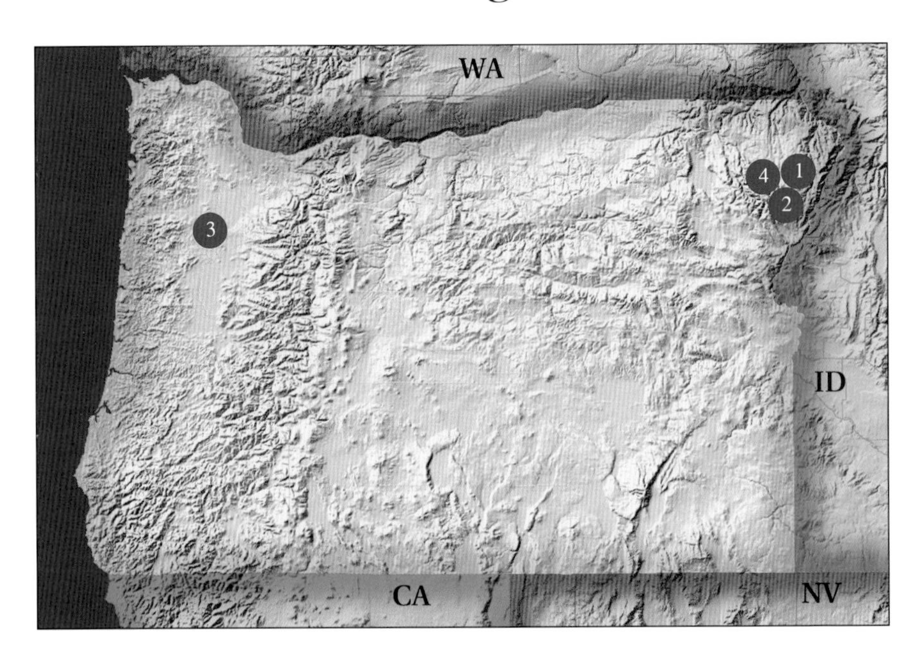

Outdoor Professionals

1. Cornucopia Wilderness Pack Station, Inc.
2. Outback Ranch Outfitters
3. S.A.L.E.M. Treks - Wiley Woods Ranch
4. Wallowa Llamas

Useful information for the state of

Oregon

State and Federal Agencies

Oregon Dept. of Fish & Wildlife
PO Box 59
Portland, OR 97207
phone: (503) 872-5268

Oregon Marine Board
435 Commercial St. NE
Salem, OR 97310
phone: (503) 373-1405
or (503) 378-8587

Columbia River Gorge Ntl. Scenic Area
902 Wasco Avenue, Ste 200
Hood River, OR 97031
phone: (541) 386-2333

Forest Service
Pacific Northwest Region
333 SW 1st Avenue
PO Box 3623
Portland, OR 97208
phone: (503) 326-2971
TTY: (503) 326-6448

Rogue River National Forest
phone: (541) 858-2200

Siskiyou National Forest
phone: (541) 471-6500

Siuslaw National Forest
phone: (541) 750-7000

Umpqua National Forest
phone: (541) 672-6601

Winema National Forest
phone: (541) 883-6714

Bureau of Land Management
Oregon State Office
(serves Washington also)
Information Access Center
1515 SW 5th Ave.
P.O. Box 2965
Portland, OR 97208-2965
phone: (503) 952-6001
or (503) 952-Plus Extension
fax: (503) 952-6308
Tdd: (503) 952-6372

Electronic mail
General Information:
or912mb@or.blm.gov
Webmaster: orwww@or.blm.gov

National Parks

Crater Lake National Park
PO Box 7
Crater Lake, OR 97604
phone: (541) 594-2211

Associations, Publications, etc.

DudeRanches.com
http://www.duderanches.com

Oregon Outdoor Association
PO Box 9486
Bend, OR 97708-9486
phone: (541) 382-9758

License and Report Requirements

- State requires licensing of Outdoor Professionals.
- State requires a "Year-End Report" for Outfitters hunting and/or fishing on BLM land.

Cornucopia Wilderness Pack Station

Eldon and Marge Deardorff
Rt. 1, Box 50 • Richland, OR 97870
phone: (541) 893-6400 • summer camp: (541) 742-5400

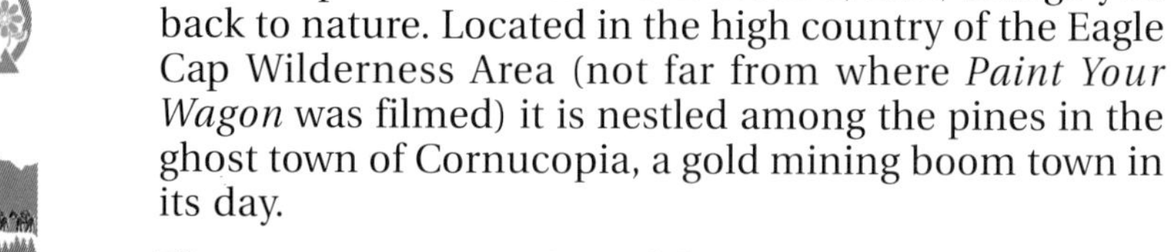

Cornucopia Wilderness Pack Station, Inc., brings you back to nature. Located in the high country of the Eagle Cap Wilderness Area (not far from where *Paint Your Wagon* was filmed) it is nestled among the pines in the ghost town of Cornucopia, a gold mining boom town in its day.

The operators are natives of the area and have complete and competent knowledge of the guide and packing business.

The pack station is used by everyone — from the avid hunter and fisherman to groups and families just wanting to get away from the hustle and bustle of the everyday life. First-timers right out of the city are in for one of the biggest thrills they will ever experience. Much of our business is derived from satisfied returning customers. We are sure you will be one of them.

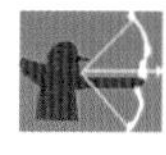

"We will give them an A+ on everything, especially the food" Mavis and Ernest Grellert

Cornucopia Wilderness Pack Trips

Outback Ranch Outfitters

Jon and Tracie Wick
P.O. Box 269 • Joseph, OR 97846
phone/fax: (541) 886-2029 • www.catsback.com/outbackranch/

Horseback Vacations

Come relax and experience the peace and tranquility of the vast wilderness. The vacation of a lifetime awaits you in one of the following wilderness areas: Eagle Cap, Snake River of Hells Canyon, or the Wenaha-Tucannon.

Horseback riding, fishing, and hunting are available in all areas.

On our most popular summer trips we spend five days flyfishing and sight-seeing the more isolated sections of the mountains — an unforgettable trip.

Just pack your clothes, sleeping bag, fishing equipment and personal gear into a duffel bag and come escape the cities and traffic jams. Leave your telephone behind and become a cowboy for a week.

Happy trails.

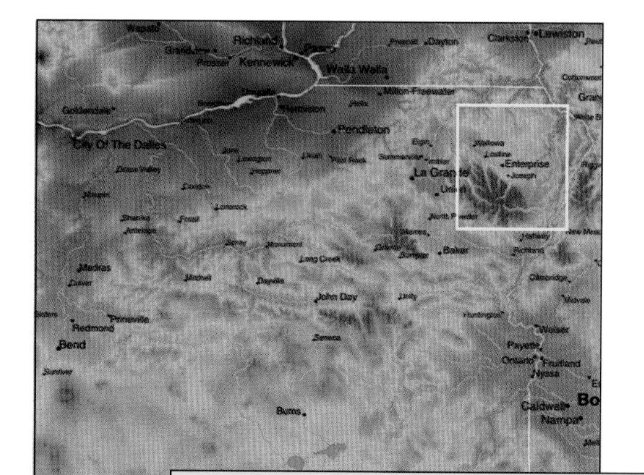

"I would highly recommend them for anyone looking for fun and adventure. One caution though....be prepared to come back every year!" Jon Peyton

Outback Ranch Outfitters

S.A.L.E.M. Treks—Wiley Woods Ranch

Ken Ploeser

555 Howell Prairie Rd. SE • Salem, OR 97301

phone/fax: (503) 362-0873 • email: MPlozr@aol.com • www.oregonlink.com/llama/

Imagine the silence in the deep forest. Gentle winds comb through 100 foot Douglas firs, a fawn dances through tall fern meadows, and you hear a soothing "hummm" from your trail companion — a llama!

Known for their ability to effortlessly carry your load, llamas are the ultimate trail companions. Youngsters and seniors, the physically challenged and hiking enthusiasts all enjoy working closely with these wonderful creatures.

Day treks are customized to meet your physical requirements and schedule. We guide you through the forest of Oregon's beautiful Silver Falls State Park, meandering through the ancient watershed of Silver Creek.

Boasting of 14 breathtaking waterfalls and the historic South Falls Lodge, Silver Falls State Park also offers full hookup and tent campsites, rental cabins, family picnic areas, horse rentals and bike trails.

"It was a lot of fun and the llamas were fun as well as funny!" Susan Mallorie

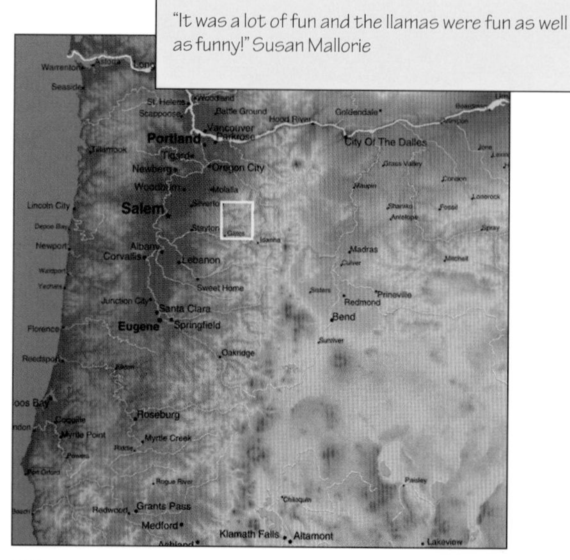

SALEM TREKS-LLAMA HIKES

Wallowa Llamas

Raz and Louise Rasmussen • Steve Backstrom

Rt. 1 Box 84 • Halfway, OR 97834

phone: (541) 742-2961 • email: wallama@pdx.oneworld.com

Since 1985, Wallowa Llamas has conducted guided tours of small groups into the state's largest wilderness area, the Eagle Cap. Here, at the southern edge of Eastern Oregon's spectacular Wallowa Mountains, amid towering peaks, glacier-sculpted valleys and sparkling mountain streams, our llamas carry the amenities, unburdening hikers to experience ease and luxury normally unavailable to backcountry travelers in such rugged environs.

Packing with even-tempered llamas, we can enjoy delightful meals prepared without freeze-dried ingredients. Wallowa Llamas provides tents, eating utensils, and all meals, beginning with lunch the first day and ending with lunch the last day.

The llamas will carry up to 20 pounds of each guest's personal gear. Anything above that must be carried by the guest. A daypack is highly recommended for carrying cameras, binoculars or rain gear.

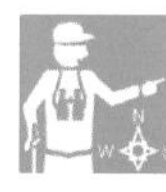

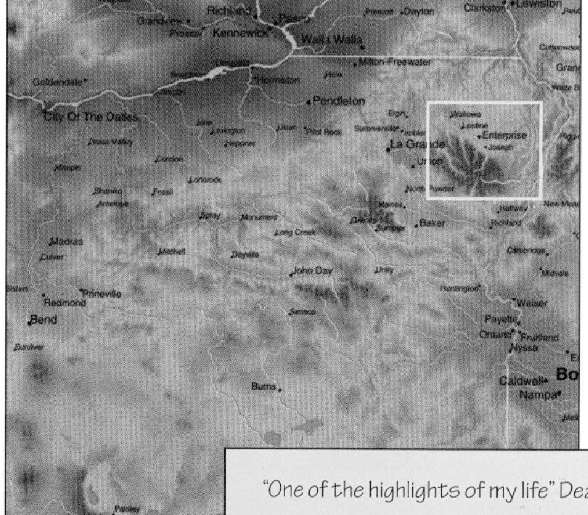

"One of the highlights of my life" Deanna Watkins

Picked-By-You Professionals in

Wyoming

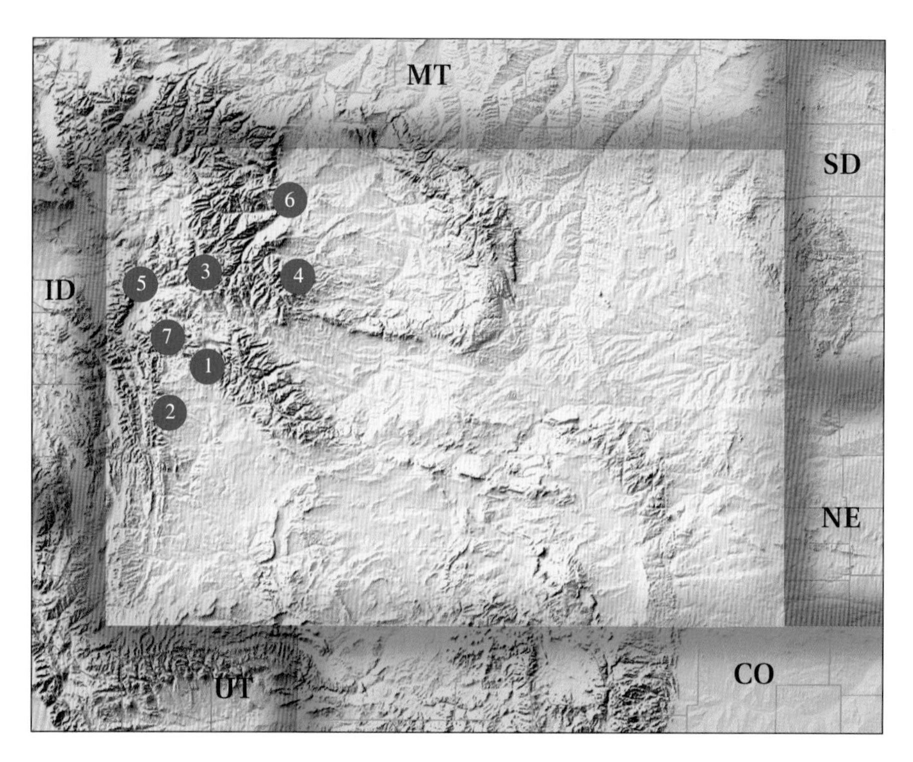

Outdoor Professionals

1. Boulder Lake Lodge
2. Darby Mountain Outfitters
3. Darwin Ranch
4. Early Guest Ranch
5. John Henry Lee Outfitters, Inc.
6. K Bar Z Guest Ranch & Outfitters
7. Lozier's Box "R" Ranch

Useful information for the state of

Wyoming

State and Federal Agencies

Wyoming Dept. of Commerce
Board of Outfitters
1750 Westland Rd.
Cheyenne, WY 82002
phone: (800) 264-0981
phone: (307) 777-5323
fax: (307) 777-6715

Wyoming Game & Fish Dept.
5400 Bishop Blvd.
Cheyenne, WY 82002
phone: (307) 777-4601

Forest Service
Intermountain Region
Federal Building
324 25th Street
Ogden, UT 84401-2310
phone: (801) 625-5306
TTY: (801) 625-5307

Bridger-Teton National Forests
Forest Service Building
340 North Cache
PO Box 1888
Jackson, WY 83001
phone: (307) 739-5500
TTY: (307) 739-5064

Bureau of Land Management
Wyoming State Office
5353 Yellowstone
P.O. Box 1828
Cheyenne, WY 82003
phone: (307) 775-6BLM or 6256
fax: (307) 775-6082

National Parks

Grand Teton National Park
PO Drawer 170
Moose, WY 83012
phone: (307) 739-3610

Yellowstone National Park
PO Box 168
Yellowstone National Park, WY 82190
phone: (307) 344-7381

Associations, Publications, etc.

DudeRanches.com
http://www.duderanches.com

Jackson Hole Chamber of Commerce
PO Box E
Jackson Hole, WY 83001
phone: (307) 733-3316
jhchamber@sisna.com

Jackson Hole Mountain Resort
Teton Village, WY 83025
phone: (800) 443-6931
info@jacksonhole.com

Wyoming Outfitters & Guides Assoc.
PO Box 2284
239 Yellowstone Ave., Suite C
Cody, WY 82414
phone: (307) 527-7453
fax: (307) 587-8633

Jackson Hole Outfitters & Guide Association
850 W. Broadway
Jackson Hole, WY 83001
phone: (307) 734-9025

License and Report Requirements

- State requires licensing of Outdoor Professionals.
- State requires that Big Game Outfitters file a “Year-End Report”.
- Fishing Outfitters need to get a permit to fish on BLM land.
- Outfitters and Guest/Dude Ranches must file a “Use” or “Day Report” with the Wyoming Forest Service if they Fish, Hunt or Raft on Forest Service Land.

Boulder Lake Lodge

Kim Bright
Box 1100H • Pinedale, WY 82941
phone: (800) 788-5401 • (307) 537-5400

Boulder Lake Lodge, located on the west slope of the remote and rugged Wind River Mountain Range, serves as headquarters for our many varied pack trip operations.

We offer guided trips ranging from hourly rides out of our rustic lodge to ten-day excursions high into the Bridger Wilderness Area. There are no roads!

Trails wind through some of the most spectacular high mountain scenery in this country at around 10,000 feet elevation. We pride ourselves in our fine horses and mules. We have a clean, professionally-staffed lodge and camp.

Exclusive groups with four or more.

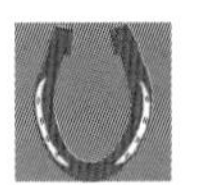

"The pleasant family atmosphere and delicious meals at this secluded ranch were delightful and the professional handling of our spot pack into the surrounding mountain was excellent!" Cynthia Fisher

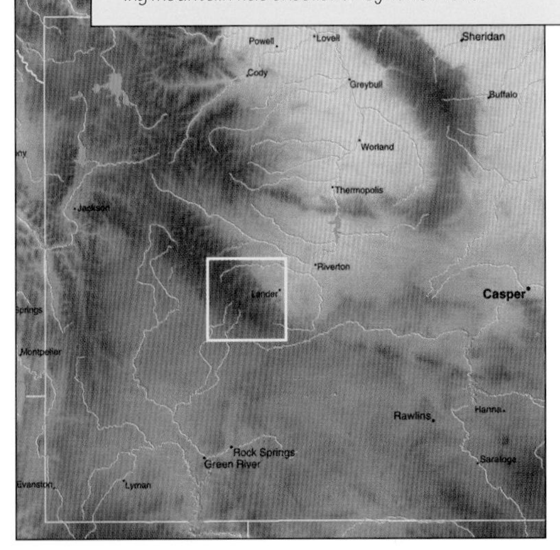

BOULDER
LAKE
LODGE

Darby Mountain Outfitters

R. John Harper and Chuck Thornton

P.O. Box 447 • Big Piney, WY 83113

phone: R.John Harper (307) 276-3934 • Chuck Thornton (307) 386-9220

Pack trips are our specialty. From late June through September catch the fragrances of spring, see newborn wildlife, and enjoy colorful Indian autumns.

Our pack trips travel through an isolated mountain range in the center of the Wyoming Rockies, a place few people have seen.

Enjoy majestic scenery, fish mountain lakes for cutthroat and brook trout, and eat healthy western campfire meals under our big sky.

Experienced guides/packers will fit you to a horse and teach you how to ride in the mountains.

Just bring an adventuresome spirit and we will provide the rest!

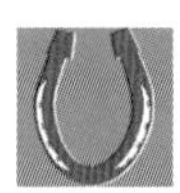

"This experience is not for the faint of heart. Chuck Thornton offers a variety of mountain trips that caters to both the novice, as well as the most experienced horseman" Richard M. Saroney

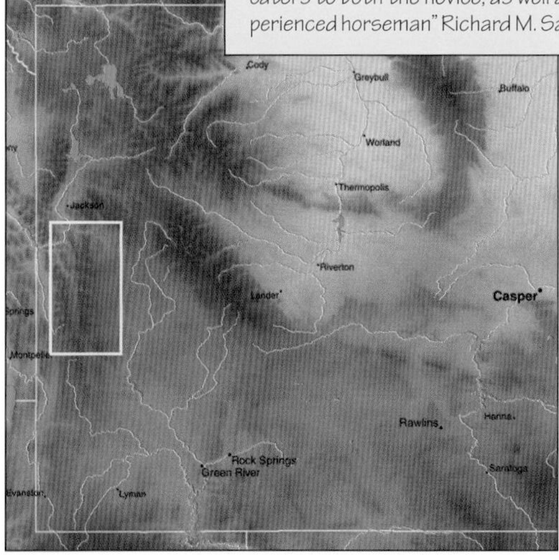

Darby Mountain Outfitters

Darwin Ranch

Loring Woodman
P.O. Box 511 • Jackson, WY 83001
phone: (307) 733-5588 • fax: (307) 739-0885

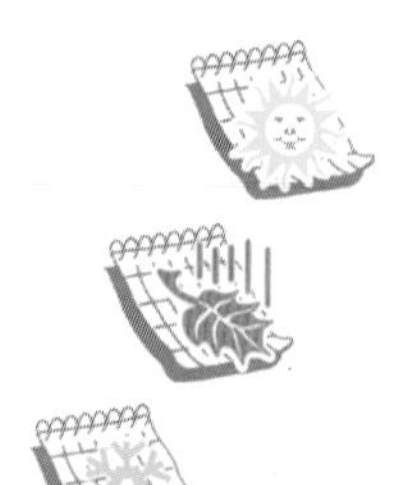

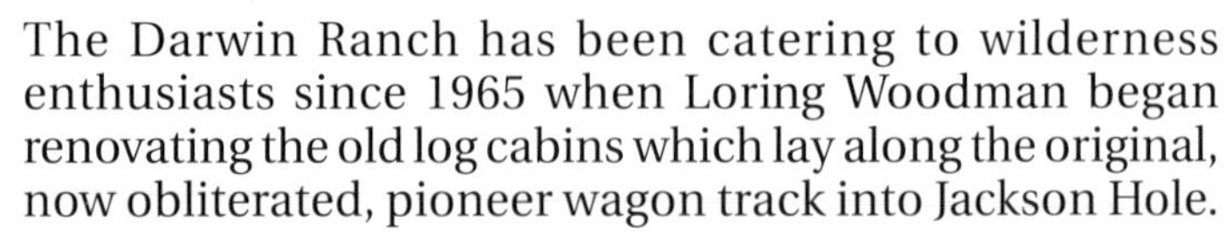

The Darwin Ranch has been catering to wilderness enthusiasts since 1965 when Loring Woodman began renovating the old log cabins which lay along the original, now obliterated, pioneer wagon track into Jackson Hole.

We are 22 miles inside Teton National Forest and have the last totally-isolated section of the Gros Ventre River to ourselves. Modern plumbing, electricity from Kinky Creek, a library, piano, and an accomplished, imaginative cook, complete the scene. Our maximum of 20 guests have a minimally-organized existence doing exactly what they want: riding, hiking, climbing, packtripping, and fly fishing.

The entire ranch, located 30 miles of Jackson Hole, is available in winter for private gatherings of six to 12.

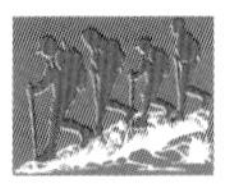

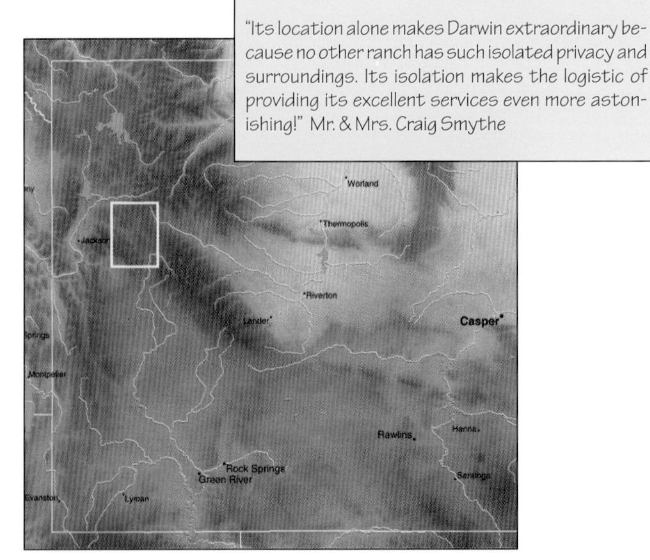

"Its location alone makes Darwin extraordinary because no other ranch has such isolated privacy and surroundings. Its isolation makes the logistic of providing its excellent services even more astonishing!" Mr. & Mrs. Craig Smythe

Early Guest Ranch

Ruth and Wayne Campbell
7374 US Hwy. 26 • Crowheart, WY 82512 (summer)
1624 Hwy. S • Wentzville, MO 63385 (winter)
phone: (307) 455-4055 (summer) (314) 332-1234 (winter) • fax: (314) 639-5250
email: earlyranch@aol.com • www.earlyranch.com

Welcome, pardner. This is where you live the cowboy way. For one unforgettable week, the spirit of the West will invade your world. Leave your phones and stress behind and escape to a world of breathtaking skies and awe-inspiring Rocky Mountain scenery.

Horseback ride to your heart's content or just kick back and relax. Daily riding instructions leads the list of exciting ranch activities, which include self-guided river trips by canoe or raft, fly or rod fishing on the gorgeous Wind River, hiking, photography, knot-tying classes, square dancing and horseshoes, evening sing-along campfires, and stargazing. Kids love our "up and at 'em" Lil' Buckaroo program.

Our unique Sunrise Spa features aerobic classes, tanning, exercise machines and hot tub. The day's itinerary is tacked to your cabin door ... do as much or as little as you care to. Boy-howdy, ya tired yet?

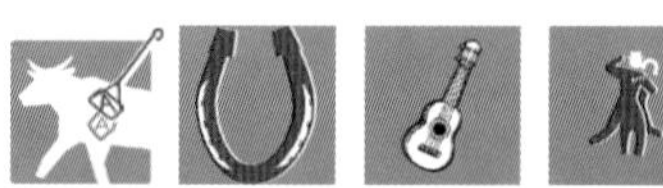

"This was truly the best vacation we ever had! This ranch was First Class all the way."
Renee Cruea

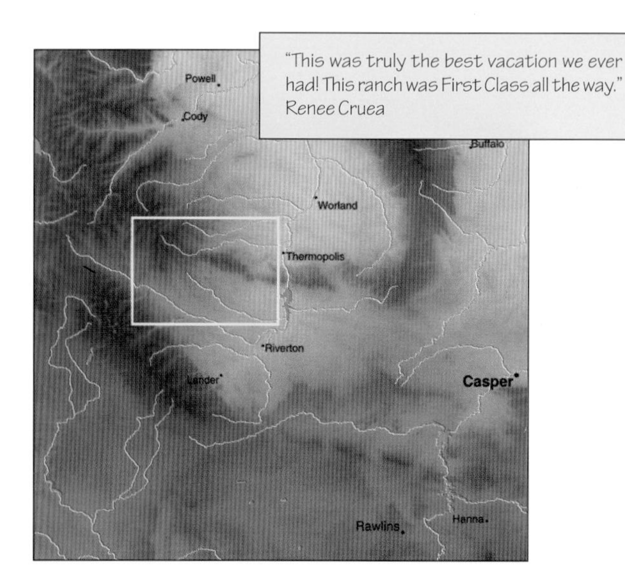

C EARLY Guest Ranch

John Henry Lee Outfitters, Inc.

John Lee
Box 8368 • Jackson, WY 83001
phone: (800) 352-2576 • (307) 733-9441 • fax: (307) 733-1403

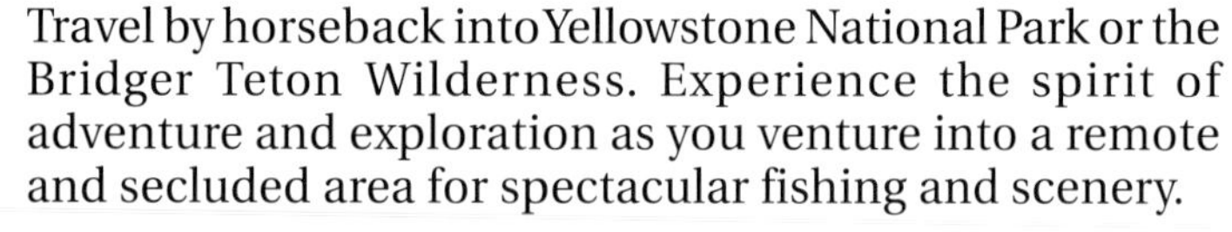

Travel by horseback into Yellowstone National Park or the Bridger Teton Wilderness. Experience the spirit of adventure and exploration as you venture into a remote and secluded area for spectacular fishing and scenery.

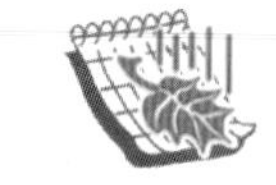

Photography and scenic trips move into the beautiful and breathtaking scenery of the high mountain country. There are abundant fields of wildflowers, lush alpine meadows, and lodgepole pine forests. This wilderness area offers you excellent opportunities to see a variety of wildlife in their natural habitat. You will enjoy the solitude and tranquil environment of the pristine wilderness.

We maintain a clean and comfortable camp. All of our equipment and horses are in excellent shape. Our mountain-wise horses can accommodate even the most novice rider.

Hearty delicious meals are served by the campfire where the stories get better and better as the stars twinkle the night away.

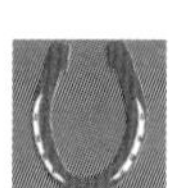

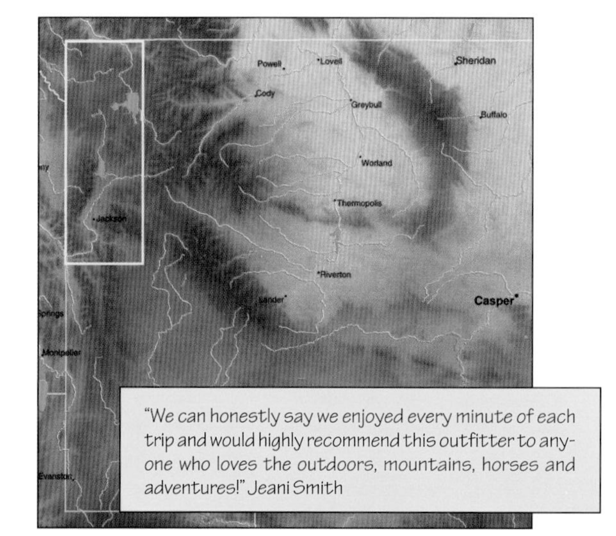

John Henry Lee
JHL
Outfitters, Inc.

K Bar Z Guest Ranch & Outfitters

Dave Segall and Dawna Barnett
P.O. Box 2167 • Cody, WY 82414
phone: (307) 587-4410 • fax: (307) 527-4605

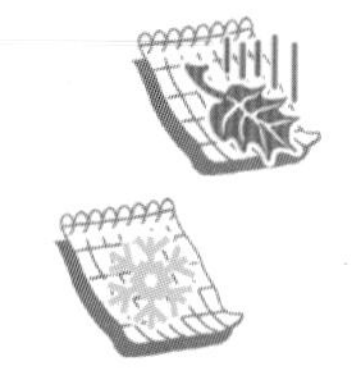

K Bar Z Guest Ranch is nestled between the Beartooth and Absaroka mountain ranges, along the Chief Joseph Scenic Highway.

This guest ranch has everything you need to make your Wyoming vacation a trip to remember. Whether it is hiking a mountain trail, horseback riding through pristine meadows or fishing the famous Clarks Fork River for native cutthroat.

For the more adventurous, pack trips into the wilderness are available. Experienced guides and gentle horses will take you into the heart of the Rockies where you can see elk, deer, moose, mountain goat and even grizzly bear.

While at the ranch you will enjoy rustic cabins, family-style meals and good old western hospitality.

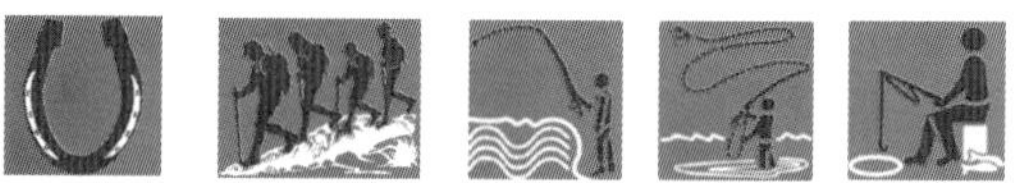

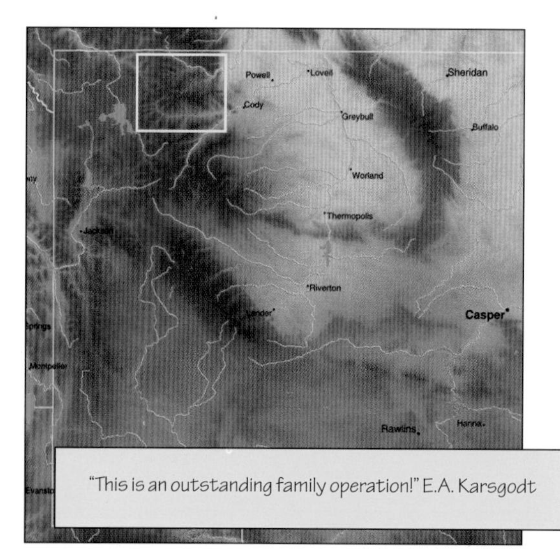

K BAR Z
GUEST RANCH
& OUTFITTERS

Lozier's Box "R" Ranch

Levi M. Lozier

Box 100-PBYB • Cora, WY 82925

phone: (800) 822-8466 • (307) 367-4868 • fax: (307) 367-6260

Experience the True West and Live that Life-Long Dream. Come join the Lozier's 100 year tradition on one of Wyoming's finest working cattle/horse guest ranches.

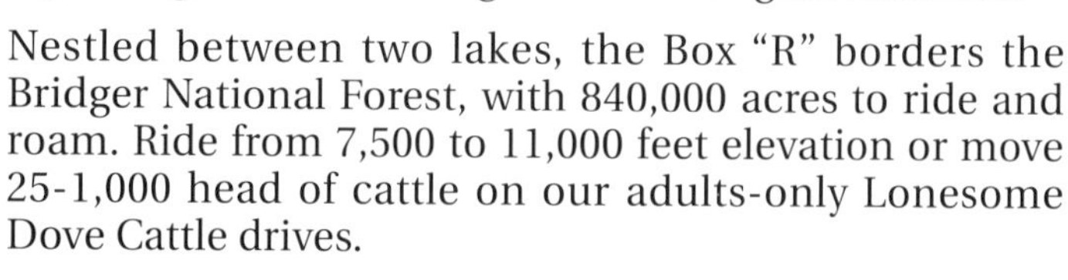

Nestled between two lakes, the Box "R" borders the Bridger National Forest, with 840,000 acres to ride and roam. Ride from 7,500 to 11,000 feet elevation or move 25-1,000 head of cattle on our adults-only Lonesome Dove Cattle drives.

To ensure you of a top-notch riding vacation, the ranch has 100 head of finely trained horses/mules and boasts a 4-to-1 horse/guest ratio.

Unlimited riding with liberties and freedoms not found on other guest ranches.

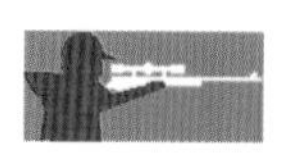

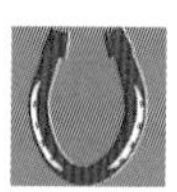

"The Lozier family is wonderful, warm, welcoming and a joy to 'reacquaint' with each year" Erin L. Burke

LOZIER'S
RANCH
CORA, WYOMING

Picked-By-You Professionals in

Canada

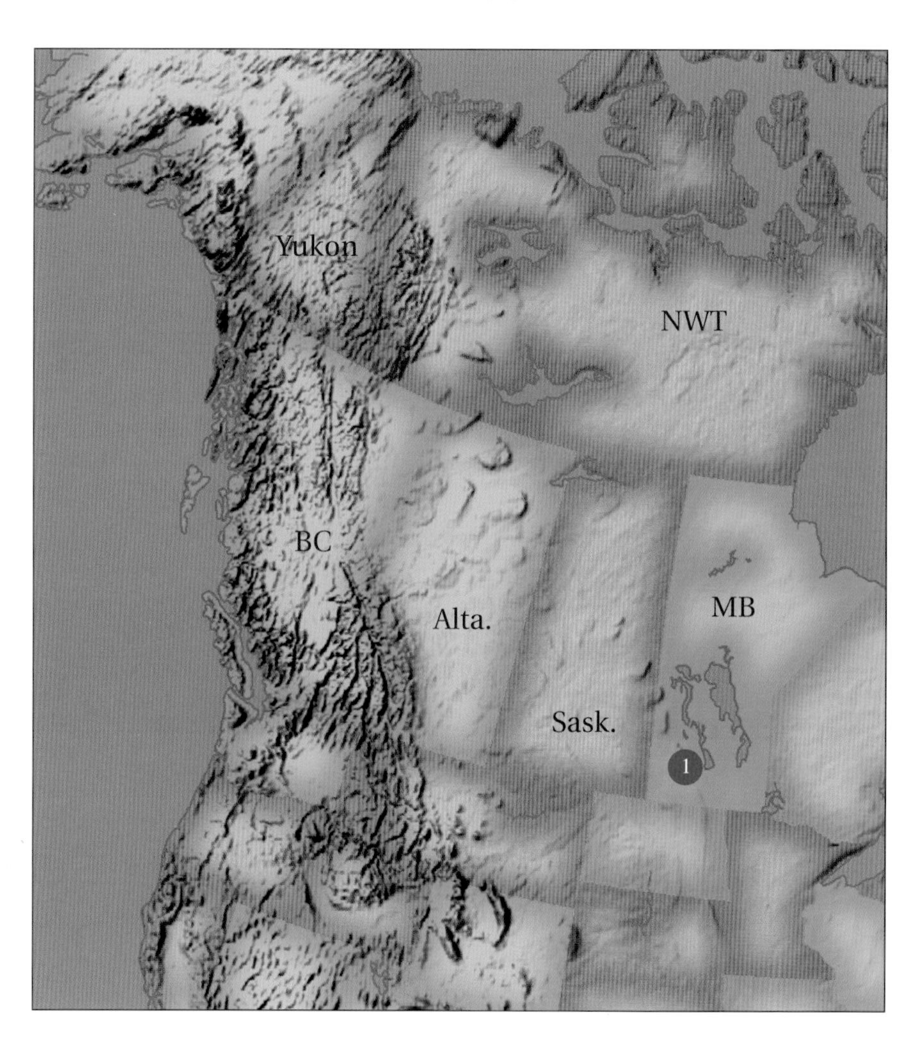

Outdoor Professionals

1 Trailhead Ranch

Useful information for the province of

Manitoba

Ministries and Agencies

Department of Natural Resources
Legislative Building, Room 333
Winnipeg, MB Canada R3C 0V8
phone: (204) 945-3730

Dept. of Industry, Trade & Tourism
Travel Manitoba, Dept RH7
1515 Carlton St.
Winnipeg, MB Canada R3C 3H8
phone: (204) 945-3777/ext. RH7
fax: (204) 945-2302

Associations, Publications, etc.

DudeRanches.com
http://www.duderanches.com

Manitoba Lodges & Outfitters Assoc.
23 Sage Crescent
Winnipeg, MB Canada R2Y 0X8
phone: (204) 889-4840

Trailhead Ranch

Anne Schuster
Box 14 • Lake Audy, Manitoba, Canada R0J 0Z0
phone: (204) 848-7649

Trailhead means "the origin of pathways to adventure." Fittingly, exciting Canadian wilderness and family adventures have been all-season services at Trailhead Ranch since 1984. Private ranchhouse accommodations are situated near wildflower meadows and aspen forests, and are flanked by 3,000 square kilometers of rugged Riding Mountain National Park wilderness.

Choose from day tours or overnight trips by covered wagon or on horseback. Hike, cycle or canoe the backcountry. At Trailhead, pioneering spirit is alive and well. "Cowboy-up!" Learn to ride, drive or pack a horse and throw a lasso. Try your hand at chuckwagon cooking. Track wildlife.

Ranch owner and active host Anne Schuster provides genuine knowledge of Canada's pioneer heritage and its wilderness legacy. Good horses, attentive hosts, and fascinating Canadian wilderness add up to great holiday adventure!

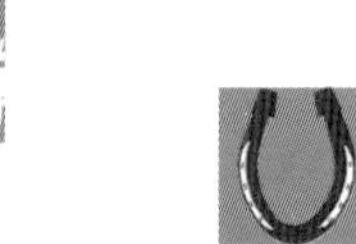

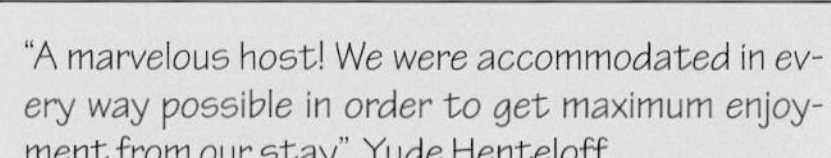

TRAILHEAD RANCH

Appendix - Western Adventures

This is a partial list of Pack Stations, Ranches Guides and Outfitters we contacted during the compilation of our book. Additional outdoor professionals and business were contacted during the compilation of Top Outfitters of Big Game Hunting. In all, during the compilation of the Picked-By-You Guides®, more than 23,000 professionals were contacted.

We invited them to participate in our survey by simply sending us their complete client list.

Some replied prizing our idea, but decided not to participate in our consumer guide books. Their main concern was the confidentiality of their client list. We truly respect this position, but we hope to have proven our honest and serious effort. We are sure they will join us in the next edition.

Others participated by sending their client list, but did not qualify for publication. In some cases because of a low score, and in other instances because of an insufficient number of questionnaires returned by their clients.

The names of the outdoor professionals who have qualified are ***italicized*** in the Appendix.

63 Ranch
Sandra M. Cahill

All 'Round Ranch

Allen's Diamond Four Wilderness Ranch
Jim Allen

Alpine House Country Chalet

Amethyst Lakes/Tonquin Valley Pack Trips
Wald & Lavone Olson

Anchor D High Mountain Hunts, Ltd.
Dewy Matthews

Anvil Butte Ranch

Bald Eagle Ranch, Inc.

Bald Mountain Outfitters, Inc.
Terry Pollard

Bar Lazy J Guest Ranch
Jerry & Cheri Helmicki

Bar M Dude Ranch
The Baker Family

Bar N Ranch

Battle Creek Lodge

Beamer's Landing/Hells Canyon Tours
Jim & Jill Koch

Bear Basin Camp
Francis Fox

Bear Creek Guest Ranch
William Beck

Bear Creek Outfitters & Guest Ranch
Bill Beck

Beartooth Ranch & JLX Outfitters
James E. Langston

Beaver Meadows Resort Ranch
North Fork Financial Corp./Don & Linda Weixelman

Benchmark Wilderness Guest Ranch

Big Hole River Outfitters
Craig Fellin

Big K Guest Ranch & Guide Service
Charles Kesterson

Big Sky Roping Ranch

Big Smoky Outfitting, Ltd.
Gary & Ricki Kruger

Big Smoky Valley Outfitters
William A. Berg

Bigfoot Mountaineering

Bill Dvorak's Kayaking & Rafting Exped.

Bill Mitchell Outfitters, Inc.
William H. & Karen Mitchell

Bitterroot Ranch
Bayard Fox

Black Mountain Outfitters

Black Otter Guide Service
Duane Neal

Blacktail Ranch
Sandra Renner

Blackwater Creek Ranch
Tom & Debbie Carlton

Bliss Creek Outfitters
Tim Doud & Doris Roesch

Blue Bronna Guiding & Outfitting
Glenn Brown

Blue Spruce Lodge & Guest Ranch

Bob's Wild West Adventures
Bob Frost

Bogie Mtn./Besa River Outfitters
Pal Gillis

Bolton Guest Ranch
Kay Bolton

Bonanza Creek Country

Boulder Lake Lodge
Kim Bright

Boulder Mountain Ranch

Boulder River Ranch

Broken Arrow Lodge
Erwin & Sherry Clark

Broken Hart Ranch
Lee I. Hart

Broken V Guest Ranch

Brooks Lake Lodge
Will Rigsby

Brush Creek Guest Ranch

Brushy Creek Guest Ranch

Buckhorn Ranch Outfitters
Harry T. Workman

Bud Nelson Outfitters
Bud Nelson

Buffalo Creek Ranch

Buffalo Horn Ranch, Inc.
James H. Walma

Buffalo Pointe, Inc.

Bull Lake Guest Ranch

Bull Mountain Outfitters
M.J. Murphy

C-B Ranch
Sandy VanderLans

C-Bar Heart Guest Ranch & Lodge

Cabinet Mountain Outfitters
Gerald & Vikki Carr

Camp Creek Inn B & B Guest Ranch

Canyon Creek Ranch

Careless Creek Getaway

Cascade Kayaks
John Amren

Castle Creek Outfitters & Guest Ranch
John D. Graham

CB Cattle & Guest Ranch

Centennial Guest Ranch

Chapoose Rivers & Trails

Cheff's Guest Ranch
Mick & Karen Cheff

Cherokee Park Ranch

Christina Falls Outfitters, Inc.
Darwin Watson

Chuck Davies Guide Service, Inc.
Mark Davies

Circle Bar Guest Ranch
Sarah Hollatz

Circle H Ranch
Pete Hegg

Circle L Ranch
Norm & Margo Leneal

Circle Z Ranch

Clear Creek Guest Ranch
Rex Frederick

Climbing Arrow Outfitters
F. & M. Anderson

Coffee Creek Ranch
Ruth Hartman

Colorado Trails Ranch

Cornucopia Wilderness Pack Station, Inc.
Eldon & Marge Deardorff

Cottonwood Ranch & Wilderness Exped.
E. Agee Smith

Cougar Country Outfitters & Guide Service
Karen Lacunate

Country Flavor
Don & Linda Kirby

Country Gardens

Country Lane Ventures
Marlene Reimer

Covered Wagon Ranch

Cow Camp

Cow Creek Ranch
Glendon & Pam Shearer

Cowboy Trails

Cowboy Village Resort

Coyote Road

Craig's Pasquia Hills Vacation Farm
Dorothy & Osborne Craig

Crazy Mountain Outfitter & Guide
Phillip Ray Keefer

Crescent H Ranch

Crossed Sabres Ranch
Fred Norris

Dalton Gang Adventures

Darby Mountain Outfitters
R.John Harper & Chuck Thornton

Darwin Ranch
Loring Woodman

Dave Flitner Packing & Outfitting
David Flitner

David Ranch
Melvin David

Daystar Guest Ranch

Dead Rock Guest Ranch

Deer Forks Ranch
Benny Middleton

Deer Springs Ranch

Deer Valley Ranch
Harold Lee DeWalt

Defa's Dude Ranch

Derringer Outfitters & Guides
David & Susan Derringer

Diamond D Ranch, Inc.
Thomas & Linda Demorest

Diamond J Guest Ranch

Diamond Jim & Sons Mountain Rides
Jim Colosimo

Diamond R Guest Ranch
James A. Slack

Dixie Outfitters, Inc.
W. Emmett & Zona Smith

Dorset Ridge Guest House

Double Diamond X Ranch
Dale Sims, Jr.

Double H Outfitters
Herb & Heather Bailey

Double K-D Ranch

Double Shot Doc Ranch

Double Spear Ranch
Tony & Donna Blackmore

DRGA Ranch

Drowsy Water Ranch
Kenneth H. Fosha

Early Guest Ranch
Ruth & Wayne Campbell

East Boulder River Guest House

East Shore Inn

Eastview Wilderness Ranch
Corrina Gray & Larry Kapeller

Eatons Ranch

Echo Canyon Guest Ranch
David & Kathleen Hampton

Ed Black Trail Rides

Ed Curnow Outfitters
Edward E. Curnow

Elk Mountain Ranch

Elkhorn Ranch
Linda Miller

Elks Hills Holiday Ranch
Leo & Erna Oestreicher

Esper's Under Wild Skies Lodge & Outfitters
Vaughn Esper

Ethilton Farms

EW Watson Watson & Sons Outfitting
Ed & Wanda Watson

Exodus Corporation
Jr. & Tony Bradbury Richard A. Bradbury

Falcon's Ledge Lodge
Altamont Flyfishers

Falcon Beach Riding Stables & Guest Ranch
Murray & Marg Imrie

Firefly Ranch
Marie Louise Link

Firehole Ranch

Fishawk River Co.
Harvey & Suzy Young

Flying A Ranch

Flying E Ranch
Vi Wellik

Flying H Ranch
John and Amee Barrus

Flying M Ranch
Barbara & Bryce Mitchell

Forsyth Ranch
Ian & Irene Forsyth

Fraser Family Farms
Ernest & Donna Fraser

Frazier Outfitters
Sammy Frazier

Frontiers North
Merv & Lynda Gunter

Fryingpan River Ranch
James B. Rea

G Bar M Ranch
Burl and LaNelle Kirkland

Grady's Farm Bed & Breakfast

Grand Canyon Bar 10 Ranch
Tony and Ruby Heaton

Granite Creek Guest Ranch
Carl & Nessie Zitlau

Grapevine Canyon Guest Ranch
Eve & Gerry Searle

Gros Ventre River Ranch
Chuck and Buzzie Smith

Haderlie's Tincup Mt. Guest Ranch
David & Lorie Haderlie

Half Moon Lake Guest Ranch
Frank Deede

Happy Hollow Vacations
Martin Capps

Hargrave Cattle & Guest Ranch
Leo & Ellen Hargrave

Hartley Guest Ranch
Doris Hartley

Hawks Point Pioneer Getaway
Don & Elaine Nielsen

Hawley Mountain Guest Ranch

Cathy Johnson

Heart Bar Ranch

Heart Six Ranch
Mike Baumann

Heaven on Earth Ranch

Hedrick Exotic Animal Farm & B&B Inn
Joe & Sondra Hedrick/Loretta Bailey

Hell Creek Guest Ranch
John E. Trumbo

Hickory Lane Farm

Hidden Creek Ranch
Iris & John Behr

Hidden Hollow Hideaway
Kelly & Jill Flynn

High Country Outfitters/Camp Wahoo
Debby Miller

High Country Vacations
Bazil Leonard & Susan Feddema

High Island Ranch & Cattle Co.
Karen Robbins

Hildreth Livestock Ranch

Hill Country Expeditions
Lois Hill

Homestead Ranch
Ed F. Arnott

Honeymoon Trail Company

Hoppe Homestead, Sinch 1866

Horse Creek Outfitters
Robert Bruce Malcolm

Horse Creek Vacation Farm
Ruby Elford

Horse Prairie Guest Ranch
Mack & Candi Hedges

Horseback Adventures Ltd.
Tom Vinson

Horseshoe Ranch
Dick Wilcox

Huckleberry Heaven Lodge

Hunewill Guest Ranch

Icicle Outfitters & Guides
Bruce & Sandy Wick

Indian Creek Guest Ranch
Mr. Bower

Iron Wheel Ranch
John & Sherry Cargill

J/L Ranch Outfitter and Guides Inc.
Linda & Joe Jessup

Jackson-Snyder Ranch

JJJ Outfitters
Max D. Barker

Joe Cantrell Outfitting
Joe Cantrell

John Henry Lee Outfitters
John Lee

Johnson Cattle Company

Jumping Rainbow Ranch

Just-N-Trails Bed & Breakfast

K Bar Z Guest Ranch & Outfitters
Dave Segall & Dawna Barnett

K Bar T Vacation Ranch
Keith & Eleanor Taylor

Kay El Bar Ranch
Jane Nash

Kedesh Guest Ranch
Charles Lander

Keenan Ranch/Bugle Ridge Outfitters

Kelan Suffold Vacation Farm
Ken & Lana Webster

Ken Sleight Pack Trips
Ken Sleight

Kennedy Meadows Resort & Pack

Keystone Resort
Phillip Stahl

Klick's K Bar L Ranch
Dick and Nancy Klick

La Gracious Stables

La Sal Mountain Guest Ranch

Lake Mancos Ranch
Todd Sehnert Kathy

Lake Upsata Guest Ranch
Richard Howe

Lakeview Resort & Outfitters
Dan & Michelle Murphy

Latigo Ranch
James A. Yost

Laughing Water Ranch
Ted & Holly Mikita-Finch

Lazy E-L Working Guest Ranch

Lazy H Guest Ranch
Karen & Phil Olbert

Lazy H Trail Co.
Richard & Connie Blair

Lazy K Bar Ranch
Barbara Van Cleve

Lazy K Bar Ranch
William Scott & Carol Moore

Lazy L & B Ranch
Lee & Bob Naylon

Let's Gallop Horeseback Adventures
Frank Deede & Kerry Thomas

Little House on the Kickapoo

Little Knife Outfitters
Glendon Nelson

Lochsa Lodge Resort

Lone Mountain Ranch
Robert L. Schaap

Lonesome Jake's Devil's Hole Ranch

Lonesome Spur Guest Ranch

Longview Farm
Bob & Charlene Siemens

Lost Creek Ranch
Les Cobb

Lost Creek Ranch Resorts
Mike & Bev Halpin

Lost Fork Ranch
Merritt G. Pride

Lost Spur Working Ranch

Lost Valley Ranch
Robert L. Foster

Lozier's Box "R" Ranch
Levi Lozier

Mackay Wilderness River Trips, Inc.
Brent Estep

Madison Valley Cabins
Gary F. Evans

Magee's Farm
Beatrice & Tom Magee

Mandorla Ranch

Marvine Ranch LLC & Elk Creek Lodge
William Wheeler

Maynard Ranch
Perry & Brenda Hunsaker

Billy & Nora Maynard

McClain Guest House

Medicine Lake Outfitters
Tom Heintz

Monture Face Outfitters
Tom Ide

Monument Canyon Ranch

Monument Valley Trail Rides

Moose Creek Ranch, Inc.
Kelly Van Orden

Moose Head Ranch
John and Eleo Mettler

Morgan Guest House

Mother Earth Lodge

Mountain Sky Guest Ranch
Shirley Arnesault

Moxie Outdoor Adventures
Cliff Stevens

Muleshoe Outfitters & Guide Service
Jack Howser

Navajo County Guided Trail Rides

Nez Perce Ranch

Nine Mile Ranch

Nine Quarter Circle Ranch
Kim & Kelly Kelsey

North Fork Ranch
Dean and Karen May

Northwest River Company
Douglas A. Tims

NOVA
Chuck Spaulding

Old Glendevey Ranch, Ltd.
Garth Peterson

Orange Torpedo Trips of Idaho
Scott Debo

Otter Basin Outfitters
Don & Beverly Gillespie

Outback Ranch Outfitters
Jon Wick

Outlaw Trails, Inc.

Pack Creek Ranch

Pack Saddle Trips

Palmquist's "The Farm"

Parade Rest Guest Ranch
Shirley Butcher

Paradise Bar Lodge
Court Boice

Paradise Guest Ranch
Jim Anderson

Passage to Utah

Peaceful Valley Lodge & Guest Ranch
Randy and Debbie Eubanks

Pepperbox Ranch

Pine Butte Guest Ranch

Pinegrove Dude Ranch
David O'Halloran

Pinehaven Bed & Breakfast

Pines Ranch Partnership
Dean Rusk & Richard Steamer

Pippin Plantation

Pleasant Vista Farm
George & Doris Husband

Powderhorn Guest Ranch
Jim & Bonnie Cook

Prairie Outfitters
Potter & Westin & Slabik

Quarter Circle "E" Outfitters & Guest
Gail Eldridge

Quarter Circle E Guest Ranch

R Lazy S Ranch
Bob & CLaire McConaughty

Rainy Hollow Wilderness Adventures Ltd.

Range Riders Ranch
Terry & Wyoma Terland

Real Ranch Living
Terry & Laurie Goehring

Red Bud Country Inn

Red Mountain Outfitters
Jim Flynn

Red Rock Adventure

Red Rock Outfitters

Red Rock Ranch
Chris and Trish Martin

Redd Ranches
David Redd

Redd Ranches Guide & Outfitter
Paul David Redd

Redfish Lake Lodge
Jack See

Redstone Trophy Hunts
David & Carol Dutchik

Reid Ranch
Mervin and Ethna Reid

Rewah Ranch
Pete Kunz

Rich Ranch
Jack & Belinda Rich

Rider Ranch

Ridge Runner Outfitters
Chad Christopherson

Riding Mountain Guest Ranch
Jim & Candy Irwin

Rimrock Ranch
Glenn Fales

River Ridge Stock & Fiber Farm

RJR Ranch

Robert Dupea Outfitters
Robert L. Dupea

Rock Creek Guest Ranch
Gayle Gibbs

Rock Creek Ranch

Rock Creek Ranch, Corp.

Rock Springs Guest Ranch
John Gill

Rocking "R" Ranch

Rocky C Adventures

Rocky Meadow Adventures

Rocky Mountain Horseback Vacations

Rogue River Raft Trips
Michelle Hanten

Ron Loucks Outfitting
Ron Loucks

Rose Ranch

Ruby's Outlaw Trail Rides

Ruby Ranch

Running-R Guest Ranch, Inc.
Ralph & Iris Kirchner & Charles "Doo" Robbins

Saddle Springs Trophy Outfitters
Bruce Cole

S.A.L.E.M. Treks-Wiley Woods Ranch
Ken Ploesser

Salmon River Challenge, Inc.
Patrick L. Marek

Salmon River Lodge
Janice Balluta

Salmon River Outfitters
Steven W. Shephard

San Juan Outfitting
Tom & Cheri Van Soelen

San Rafael Trail Rides

Scenic Rim Trail Rides

Scenic Safaris

Schively Ranch

Schmittel Packing & Outfitting
David & Verna Schmittel

Schwab Simmentals
David & Janet Schwab

Scott Valley Resort & Guest Ranch
om & Kathleen Cooper

Seton Trails Guest Lodge
De'Athe Family

Seven D Ranch
Nikki & Marshall Dominick I. Ward

Seven Devils Ranch
Rich & Judy Cook

Sheep Mountain Outfitters
Tim Haberberger

Shepp Ranch Idaho
Virginia Hopfenbeck

Shores Outfitting
Eric R. Shores

Silverwolf Chalet Resort
James & Bonnie Kennedy

Sky Corral Guest Ranch
David Vannice

Skyline Guest Ranch
David Farny

Skyline Trail Rides
Dave Flato

Skyview Ranch

Small Cattle Company
Butch Small

Snake River Park, Inc.
Stan & Karen Chatham

South Ram Outfitters
Lorne Hindbo

Spanish Springs Ranch
Jim Vondracek & Sharon Roberts

Spirit Lake Lodge
George & Connie Coonradt

Spotted Bear Ranch
Kirk & Cathy Gentry

Spotted Horse Ranch
Dick Bess

Spring Valley Guest Ranch
Jim Saville

Spur Cross Ranch
Arnold Ginsberg

Stanley Potts Outfitters
Stan & Joy Potts

Steward Ranch Outfitters
Laverne Gwaltney

Sulphur Creek Ranch
Tom T. Allegrezza

Sundance Ranch

Sunset Guest Ranch
Mike C. McCormick

Sunset Guiding & Outfitting
Duane D. Papke

Sunshine Valley Guest Ranch

Sweet Grass Ranch
Bill and Shelly Carroccia

Sylvan Dale Ranch
Susan and David

Sylvania Tree Farm

T Cross Ranch
Ken & Garey Neal

T Lazy B Ranch
Robert L. Walker

Tanner Ranch

Tawanta Ranch
Monte Farnsworth

Teton Ridge Guest Ranch
Albert Tilt, III

The Alternate Transit Authority

The Big K Guest Ranch Kathie Williamson

The Don K Ranch

The Elkhorn Guest Ranch

The Hideout, Flitner Ranch
Kathryn Flitner

The Hitching Post

The Home Ranch
Ken & Cile Jones

The Lodge at Chama
Frank Simms

The Rising Wolf Ranch

The Seventh Ranch

The Stoney Lonesome Ranch

The Stopping Place
Deb & Ian McLeod & Family

The Wald Ranch

Three Bars Cattle & Guest Ranch
Jeff & April Beckley

Three Rivers Resort & Campground
George Michael Smith

Thunder Mountain Outfitters
Cameron Garnick

Thunder Valley Bed & Breakfast

Torch Valley Country Retreats
George & Jean Lidster

Totem Pole Tours & Trailrides

Toussaint Ranch

Trail-Side Bed 'n Breakfast

Trail Creek Ranch
Elizabeth Woolsey & Alexandra Grant

Trail Farm

Trailhead Ranch
Anne Schuster

Triangle C Ranches
Ron Gillett

Triangle X Ranch
Donald Turner

Trillium

Triple Creek Ranch

Triple R Ranch
Jack and Cherrylee Bradt

Tuck-A-Way Camp

Turpin Meadow Ranch
Stan Castagno

Twin Peaks Ranch Inc.
Dave Giles

Two Bars Seven Ranch

Two Spirit Guest Ranch & Retreat
Lee Cryer & Denise Needham

U-Bar Wilderness Ranch

Utah Escapades

UXU Ranch
Hamilton Bryan

Vanhaur Polled Hereford Ranch

Vermejo Park Ranch
Jim Baker

Vista Verde Guest Ranch
ohn S. Munn

Wagons Roll Inn
Lorraine & Ray Ostrom

Wahoos Adventures TN, Inc.
Melissa France

Wallowa Llamas
Raz & Louise Rasmussen, Steve Backstrom

Wapiti Meadow Ranch & Outfitters
M. Barry Bryant & Diane Haynes

Waunita Hot Springs Ranch
Ryan Pringle

West Fork Meadows Ranch, Inc.
Guido Oberdorfer

Western Adventures
Glenn & Leslie Huber

Western Pleasure Guest Ranch

White Buffalo Ranch Retreat

White Pine Ranch
Dennis & Cindy Hall

White Stallion Ranch
Russell True

White Tail Ranch/WTR Outfitters, Inc.
Jack & Karen Hooker

Whitewater Wilderness Lodge

Why Lazy Tee Ranch

Wickens Salt Creek Ranch

Wild Horse Creek Ranch
William R. Shields & Rick Hankins

Wilderness Connection, Inc.
Charles G. Duffy

Wilderness Outfitters
Shelda, Justin & Jarrod Farr Scott

Wilderness Trails Ranch
Gene Roberts

Wildlife Adventures, Inc.
Jack E. Wemple

Wind River Wilderness Tours

Wind Valley Guiding & Outfitting
Ken Fraser & Shelly Paul

Windy Acres Vacation Farm
Elliot & Reta Kimpton

Wit's End Guest Ranch & Resort
Jim & Lynn Custer

Wolf Creek Outfitters, LLC
Jason Ward

Wolf Mountain Ranch

Wolfway Farm

Woodside Ranch Resort
William Feldmann

Worldwide Outdoor Adventures
Randy Beck

Wounded Knee
Dick & Judy Wells

Wright's Vacation Farm
Ken & Linda Wright

Yellowstone Valley Ranch

ZAK Inn Guest Ranch

Zion's Kolob Mountain Ranch

Zion's Ponderosa Resort

Peel off your address label and place here

PLEASE NOTE: The Business or Professional you are invited to rate is printed above your name after "referred by..."

Picked-By-You Questionnaire
Top Western Adventures

Name of the Ranch:____

Name of your Host____________________ Length of your Stay____________

Date of Trip______________ Location_______________________

Day Ride ☐ Pack Trip ☐ Cattle Drive ☐ Guest Ranch ☐ Other________

Was this a Family Trip where your children were actively involved in the activities? Yes ☐ No ☐

	Question	Outstanding	Excellent	Good	Acceptable	Poor/Inferior	Unacceptable
1.	How helpful was the Outfitter (Ranch or Pack Station) with travel arrangements, dates, special accommodations, etc.?...................................	☐	☐	☐	☐	☐	☐
2.	How well did the Outfitter (Ranch or Pack Station) provide important details that better prepared you for your experience (clothing, list of "take along", etc.)?..	☐	☐	☐	☐	☐	☐
3.	How would you rate the Outfitter's (Ranch or Pack Station) office skills in handling deposits, charges, reservations, returning calls before and after your trip?...	☐	☐	☐	☐	☐	☐
4.	How would you rate the accommodations (bunk house, tent, cabin, lodge, etc.)?..	☐	☐	☐	☐	☐	☐
5.	How would you rate the equipment provided by the Outfitter (Ranch or Pack Station) (horses, saddles, wagons, pick-ups, etc.)?................................	☐	☐	☐	☐	☐	☐
6.	How would you rate the cooking (quantity, quality and cleanliness of the service)?..	☐	☐	☐	☐	☐	☐
7.	How would you rate your Guide/Host's attitude..	☐	☐	☐	☐	☐	☐
8.	How would you rate your Guide/Host's professionalism?............................	☐	☐	☐	☐	☐	☐
9.	How would you rate your Guide/Host's disposition?....................................	☐	☐	☐	☐	☐	☐
10.	How would you rate your Guide/Host's knowledge of the area?..................	☐	☐	☐	☐	☐	☐

	Outstanding	Excellent	Good	Acceptable	Poor/Inferior	Unacceptable
11. How would you rate your Guide/Host's knowledge of the livestock?...........	☐	☐	☐	☐	☐	☐
12. How would you rate your Guide/Host's handling of the livestock?...............	☐	☐	☐	☐	☐	☐
13. How would you rate the quality of the different activities offered during your stay (roping, branding, horse care, ranch life, etc.)?................................	☐	☐	☐	☐	☐	☐
14. How flexible was your Guide or Host in trying to meet your goal(s)?...........	☐	☐	☐	☐	☐	☐
15. How would you rate the overall quality of your outdoor experience?..........	☐	☐	☐	☐	☐	☐

	Good	Fair	Poor
16. How would you describe the weather conditions?...	☐	☐	☐

17. Was the overall quality of your experience (quality of animals, activities, and accommodations, etc.) accurately represented by the Outfitter (Ranch or Pack Station)? .... ☐ Yes ☐ No

18. Did you provide the Outfitter (Ranch or Pack Station) with truthful statements regarding your personal needs, your skills and your expectations?... ☐ Yes ☐ No

19. Would you use this Outdoor Professional/Business again?........................... ☐ Yes ☐ No

20. Would you recommend this Outdoor Professional/Business to others?..... ☐ Yes ☐ No

Comments:____

Will you permit Picked-By-You to use your name and comments in our book(s)? ☐ Yes ☐ No

Signature______________________________

Cross Index by Activity

Archery

Coffee Creek Ranch
CornucopiaWilderness Pack Stn.
Esper's UnderWild Skies
EW Watson & Sons
Echo Canyon Ranch
Granite Creek Guest Ranch
IronWheel Ranch
Maynard Ranch
Outback Ranch Outfitters
Spanish Spring Ranch

Backpack/Trekking Excursions

Skyline Guest Ranch
Wallowa Llamas

Barrel Racing

Clear Creek Ranch
Coffee Creek Ranch
Early Guest Ranch
Hidden Hollow Hideway
Spanish Spring Ranch
Trailhead Ranch

Big Game Hunting

Boulder Lake Lodge
Cheff Guest Ranch
Coffee Creek Ranch
CornucopiaWilderness Pack Stn.
Darby Mountain Outfitters
Darwin Ranch
Echo Canyon
Esper's UnderWild Skies
EW Watson & Sons
Frazier Outfitting
K Bar Z Guest Ranch
Hidden Hollow Hideway
IronWheel Ranch
John Henry Lee Outfitters
Lakeview Resort
Lost Creek Ranch
Lozier's Box "R" Ranch
Monture Face Outfitters
Outback Ranch Outfitters
San Juan Outfitting
WhiteTail Ranch/WTR Outf.

Branding

Cheff Guest Ranch
Double Spear Ranch
Early Guest Ranch
Echo Canyon Ranch
EW Watson & Sons
Hargrave Cattle & Guest Ranch
Hartley Guest Ranch
Hidden Hollow Hideway
Lozier's Box "R" Ranch
Maynard Ranch
Outback Ranch Outfitters
Spanish Spring Ranch

Cattle and Horse Drives

Cheff Guest Ranch
Double Spear Ranch
Echo Canyon Ranch
EW Watson & Sons
Granite Creek Guest Ranch
Hargrave Cattle & Guest Ranch
Hartley Guest Ranch
Hidden Hollow Hideway
Little Knife Outfitters
Lozier's Box "R" Ranch
Maynard Ranch
Outback Ranch Outfitters

Cross Index by Activity

Children/Youth

Clear Creek Ranch
Coffee Creek Ranch
Granite Creek Guest Ranch
Iron Wheel Ranch
Maynard Ranch
Nine Quarter Circle Ranch
Rich Ranch
Skyline Guest Ranch

Cowboy and Live Entertainment

Coffee Creek Ranch
Early Guest Ranch
Echo Canyon Guest Ranch
Granite Creek Guest Ranch
Hargrave Cattle & Guest Ranch
Maynard Ranch
Schmittel Packing and Outfitting
Skyline Guest Ranch

Clay Shooting

Echo Canyon Ranch
Hargrave Cattle & Guest Ranch
Spanish Spring Ranch

Cross Country Skiing

Beaver Meadows Resort Ranch
Coffee Creek Ranch
Cornucopia Wilderness Pack Stn.
Darwin Ranch
Hargrave Cattle & Guest Ranch
K Bar Z Guest Ranch
Rich Ranch
Skyline Guest Ranch
Spanish Spring Ranch

Cutting

Granite Creek Guest Ranch
Hargrave Cattle & Guest Ranch
Lozier's Box "R" Ranch

Dancing

Clear Creek Ranch
Early Guest Ranch
Nine Quarter Circle Ranch
Spanish Spring Ranch

Fishing

Cheff Guest Ranch
Clear Creek Ranch
Darby Mountain Outfitters
Early Guest Ranch
Echo Canyon Ranch
Esper's Under Wild Skies
Frazier Outfitting
Granite Creek Guest Ranch
Hartley Guest Ranch
Iron Wheel Ranch
K Bar Z Guest Ranch
Lakeview Resort
Maynard Ranch
Nine Quarter Circle Ranch
Outback Ranch Outfitters
San Juan Outfitting
Schmittel Packing and Outfitting
Skyline Guest Ranch
Spanish Spring Ranch
Trailhead Ranch
Wallowa Llamas
White Tail Ranch/WTR Outf.

Cross Index by Activity

Fly Fishing

Beaver Meadows Resort Ranch
Boulder Lake Lodge
Broken Arrow Lodge
Cheff Guest Ranch
Clear Creek Ranch
Coffee Creek Ranch
Cornucopia Wilderness Pack Stn.
Darby Mountain Outfitters
Darwin Ranch
Double Spear Ranch
Early Guest Ranch
Echo Canyon Ranch
Esper's Under Wild Skies
EW Watson & Sons
Frazier Outfitting
Hargrave Cattle & Guest Ranch
Hidden Hollow Hideway
Iron Wheel Ranch
K Bar Z Guest Ranch
Lakeview Resort
Lost Creek Ranch
Lozier's Box "R" Ranch
Maynard Ranch
Monture Face Outfitters
Nine Quarter Circle Ranch
Rich Ranch
San Juan Outfitting
Skyline Guest Ranch
Trailhead Ranch
Wallowa Llamas
White Tail Ranch/WTR Outf.

Float Fishing

Darby Mountain Outfitters

Golfing

Clear Creek Ranch
Echo Canyon Ranch
Trailhead Ranch
Skyline Guest Ranch

Ghost Town Tours

Early Guest Ranch
EW Watson & Sons
Iron Wheel Ranch
Skyline Guest Ranch

Hiking

Broken Arrow Lodge
Cheff Guest Ranch
Clear Creek Ranch
Cornucopia Wilderness Pack Stn.
Darby Mountain Outfitters
Darwin Ranch
Early Guest Ranch
Esper's Under Wild Skies
Frazier Outfitting
Granite Creek Guest Ranch
Hartley Guest Ranch
K Bar Z Guest Ranch
Lost Creek Ranch
Lozier's Box "R" Ranch
Maynard Ranch
Nine Quarter Circle Ranch
Outback Ranch Outfitters
S.A.L.E.M. Treks
Skyline Guest Ranch
Spanish Spring Ranch
Trailhead Ranch
Wallowa Llamas
White Tail Ranch/WTR Outf.

Cross Index by Activity

Horseback Riding

Beaver Meadows Resort Ranch
Boulder Lake Lodge
Broken Arrow Lodge
Cheff Guest Ranch
Clear Creek Ranch
Coffee Creek Ranch
Cornucopia Wilderness Pack Stn.
Darby Mountain Outfitters
Darwin Ranch
Double Spear Ranch
Early Guest Ranch
Echo Canyon Ranch
Esper's Under Wild Skies
EW Watson & Sons
Frazier Outfitting
Granite Creek Guest Ranch
Hargrave Cattle & Guest Ranch
Hartley Guest Ranch
Hidden Hollow Hideway
Iron Wheel Ranch
John Henry Lee Outfitters
K Bar Z Guest Ranch
Lakeview Resort
Little Knife Outfitters
Lost Creek Ranch
Lozier's Box "R" Ranch
Maynard Ranch
Monture Face Outfitters
Nine Quarter Circle Ranch
Rich Ranch
San Juan Outfitting
Schmittel Packing and Outfitting
Skyline Guest Ranch
Spanish Spring Ranch
Trailhead Ranch
White Tail Ranch/WTR Outf.

Horse Pack Trips

Boulder Lake Lodge
Broken Arrow Lodge
Cheff Guest Ranch
Cornucopia Wilderness Pack Stn.
Darby Mountain Outfitters
Darwin Ranch
Double Spear Ranch
Early Guest Ranch
Echo Canyon Ranch
Esper's Under Wild Skies
EW Watson & Sons
Frazier Outfitting
Hargrave Cattle & Guest Ranch
Hartley Guest Ranch
Hidden Hollow Hideway
John Henry Lee Outfitters
K Bar Z Guest Ranch
Lakeview Resort
Little Knife Outfitters
Lost Creek Ranch
Lozier's Box "R" Ranch
Maynard Ranch
Monture Face Outfitters
Nine Quarter Circle Ranch
Rich Ranch
San Juan Outfitting
Schmittel Packing and Outfitting
Skyline Guest Ranch
Trailhead Ranch
White Tail Ranch/WTR Outf.

Horseshoe

Beaver Meadows Resort Ranch
Boulder Lake Lodge
Broken Arrow Lodge
Cheff Guest Ranch
Coffee Creek Ranch
Darby Mountain Outfitters
Early Guest Ranch
Echo Canyon Ranch

Cross Index by Activity

Granite Creek Guest Ranch
Hargrave Cattle & Guest Ranch
Hidden Hollow Hideway
Iron Wheel Ranch
John Henry Lee Outfitters
K Bar Z Guest Ranch
Lozier's Box "R" Ranch
Maynard Ranch
Nine Quarter Circle Ranch
Outback Ranch Outfitters

Llama Pack Trips

S.A.L.E.M. Treks
Wallowa Llamas

Mountain Biking

Beaver Meadows Resort Ranch
Echo Canyon Ranch
Skyline Guest Ranch
Trailhead Ranch

Raft / Float Trips

Clear Creek Ranch
Cornucopia Wilderness Pack Stn.
Early Guest Ranch
Echo Canyon Ranch
EW Watson & Sons
Hargrave Cattle & Guest Ranch
Hartley Guest Ranch
Iron Wheel Ranch

Roping

Clear Creek Ranch
Double Spear Ranch
Echo Canyon Ranch
EW Watson & Sons
Granite Creek Guest Ranch
Hargrave Cattle & Guest Ranch
Hidden Hollow Hideway
Lozier's Box "R" Ranch
Trailhead Ranch

School, Fly Fishing

Skyline Guest Ranch

School, Riding

Beaver Meadows Resort Ranch
Broken Arrow Lodge
Cheff Guest Ranch
Darby Mountain Outfitters
Double Spear Ranch
Early Guest Ranch
Echo Canyon Ranch
Hargrave Cattle & Guest Ranch
Iron Wheel Ranch
John Henry Lee Outfitters
Nine Quarter Circle Ranch
Skyline Guest Ranch
Spanish Spring Ranch

Cross Index by Activity

School, Horse Packing

Boulder Lake Lodge
IronWheel Ranch
Hidden Hollow Hideway
Maynard Ranch
Trailhead Ranch

School, Horse Shoeing

IronWheel Ranch

Snowshoeing

Beaver Meadows Resort Ranch
Coffee Creek Ranch
Skyline Guest Ranch

Target Shooting

Broken Arrow Lodge
Coffee Creek Ranch
Double Spear Ranch
Echo Canyon Ranch
Esper's Under Wild Skies
EW Watson & Sons
Granite Creek Guest Ranch
Hargrave Cattle & Guest Ranch
Hidden Hollow Hideway
IronWheel Ranch
Lost Creek Ranch
Lozier's Box "R" Ranch
Maynard Ranch
Spanish Spring Ranch

Team Penning

Granite Creek Guest Ranch
Hargrave Cattle & Guest Ranch

Team Roping

Early Guest Ranch

Wagon Rides

Beaver Meadows Resort Ranch
Coffee Creek Ranch
CornucopiaWilderness Pack Stn.
Echo Canyon Ranch
EW Watson & Sons
Lozier's Box "R" Ranch
Maynard Ranch
Nine Quarter Circle Ranch
Spanish Spring Ranch
Trailhead Ranch

Cross Index by Activity

Wildlife Viewing

Index of Ranches, Outfitters & Pack Stations by State/Province

Index of Ranches, Outfitters & Pack Stations by State/Province

Photo Credits

Double Spear Ranch/©Sorrel 71
Nine Quarter Circle Ranch/©Dann Coffey 85

Alphabetical Index by Company Name